Artificial Intelligence Techniques in Power Systems Operations and Analysis

An electrical power system consists of a large number of generation, transmission, and distribution subsystems. It is a very large and complex system; hence, its installation and management are very difficult tasks. An electrical system is essentially a very large network with very large data sets. Handling these data sets can require much time to analyze and subsequently implement. An electrical system is necessary but also potentially very dangerous if not operated and controlled properly. The demand for electricity is ever increasing, so maintaining load demand without overloading the system poses challenges and difficulties.

Thus, planning, installing, operating, and controlling such a large system requires new technology. Artificial intelligence (AI) applications have many key features that can support a power system and handle overall power system operations. AI-based applications can manage the large data sets related to a power system. They can also help design power plants, model installation layouts, optimize load dispatch, and quickly respond to control apparatus. These applications and their techniques have been successful in many areas of power system engineering.

Artificial Intelligence Techniques in Power Systems Operations and Analysis focuses on the various challenges arising in power systems and how AI techniques help to overcome these challenges. It examines important areas of power system analysis and the implementation of AI-driven analysis techniques. The book helps academicians and researchers understand how AI can be used for more efficient operation. Multiple AI techniques and their application are explained. Also featured are relevant data sets and case studies.

Highlights include:

- Power quality enhancement by PV-UPQC for non-linear load
- Energy management of a nanogrid through flair of deep learning from IoT environments
- Role of artificial intelligence and machine learning in power systems with fault detection and diagnosis
- AC power optimization techniques
- Artificial intelligence and machine learning techniques in power systems automation

ADVANCES IN COMPUTATIONAL COLLECTIVE INTELLIGENCE
Edited by
Dr. Subhendu Kumar Pani
Principal, Krupajal Group of Institutions, India

Published

Applications of Machine Learning and Deep Learning on Biological Data
By Faheem Syeed Masoodi, Mohammad Tabrez Quasim, Syed Nisar Hussain Bukhari, Sarvottam Dixit, and Shadab Alam
ISBN: 978-1-032-214375

Artificial Intelligence Techniques in Power Systems Operations and Analysis
By Nagendra Singh, Sitendra Tamrakar, Arvind Mewara, and Sanjeev Kumar Gupta
ISBN: 978-1-032-294865

Technologies for Sustainable Global Higher Education
By Maria José Sousa, Andreia de Bem Machado, and Gertrudes Aparecida Dandolini
ISBN: 978-1-032-262895

Forthcoming

Artificial Intelligence and Machine Learning for Risk Management of Natural Hazards and Disasters
By Cees van Westen, Romulus Costache, Dimitrios A. Karras, R. S. Ajin, and Sekhar L. Kuriakose
ISBN: 978-1-032-232768

Computational Intelligence in Industry 4.0 and 5.0 Applications: Challenges and Future Prospects
Joseph Bamidele Awotunde, Kamalakanta Muduli, and Biswajit Brahma
ISBN: 978-1-032-539225

Deep Learning for Smart Healthcare: Trends, Challenges and Applications
K. Murugeswari, B. Sundaravadivazhagan, S. Poonkuntran, and Thendral Puyalnithi
ISBN: 978-1-032-455815

Edge Computational Intelligence for AI-Enabled IoT Systems
By Shrikaant Kulkarni, Jaiprakash Narain Dwivedi, Dinda Pramanta, and Yuichiro Tanaka
ISBN: 978-1-032-207667

Explainable AI and Cybersecurity
By Mohammad Tabrez Quasim, Abdullah Alharthi, Ali Alqazzaz, Mohammed Mujib Alshahrani, Ali Falh Alshahrani, and Mohammad Ayoub Khan
ISBN: 978-1-032-422213

https://www.routledge.com/Advances-in-Computational-Collective-Intelligence/book-series/ACCICRC

Artificial Intelligence Techniques in Power Systems Operations and Analysis

Edited by
Dr. Nagendra Singh
Dr. Sitendra Tamrakar
Arvind Mewada
Dr. Sanjeev Kumar Gupta

CRC Press
Taylor & Francis Group
Boca Raton London New York

CRC Press is an imprint of the
Taylor & Francis Group, an **informa** business
AN AUERBACH BOOK

First edition published 2024
by CRC Press
2385 Executive Center Drive, Suite 320, Boca Raton, FL 33431

and by CRC Press
4 Park Square, Milton Park, Abingdon, Oxon, OX14 4RN

CRC Press is an imprint of Taylor & Francis Group, LLC

© 2024 Taylor & Francis Group, LLC

ISBN: 978-1-032-29486-5 (hbk)
ISBN: 978-1-032-29492-6 (pbk)
ISBN: 978-1-003-30182-0 (ebk)

DOI: 10.1201/9781003301820

Typeset in Garamond
by Apex CoVantage, LLC

Contents

Contributors

Ayush Agrawal
Computer Science and Engineering
  Department
Motilal National Institute of
  Technology Allahabad
Prayagraj, India

Rahul Agrawal
Department of Electrical Engineering,
Guru Gobind Singh College of
  Engineering and Research Centre
Nashik, India

Shekh Kulsum Almas
Sage University
Indore, India

Eknath Borkar
Department of Electrical and Electronics
  Engineering
Technocrats Institute of
  Technology
Bhopal. Madhya Pradesh, India

Ravi Kant Choubey
Amity Institute of Applied Sciences
Amity University
Noida, Uttar Pradesh, India

Rupesh Kumar Dewang
Computer Science and Engineering
  Department
Motilal National Institute of
  Technology Allahabad
Prayagraj, India

Devashish Dhaulakhandi
School of Computer Science and
  Engineering
VIT Bhopal University
Sehore, India

Kaustubh Dwivedi
UIT Polytechnic
Rajiv Gandhi Proudyogiki Vishwavidyalaya
Bhopal, India

Swati Gade
Department of Electrical Engineering
Sandip Institute of Technology and
  Research Centre
Nashik, India

Shivani Gautam
Chitkara University Institute of
  Engineering and Technology
Chitkara University
Himachal Pradesh, India

Amit Gupta
Department of Electrical and Electronics
  Engineering
Gyanganga Institute of Science and
  Technology
Jabalpur, Madhya Pradesh, India

Akshay Jadhav
School of Computer Science and
  Engineering
VIT Bhopal University
Sehore, India

Pradeep Kumar Jhinge
Jabalpur Engineering College
Jabalpur, Madhya Pradesh, India

Rachana Kamble
Department of CSE
Technocrats Institute of Technology
  Excellence
Bhopal, Madhya Pradesh, India

Hameed Khan
Jabalpur Engineering College
Jabalpur, Madhya Pradesh, India

Kamal Kumar Kushwah
Jabalpur Engineering College
Jabalpur, Madhya Pradesh, India

Rupesh Kushwah
Govt. SSA, PG College
Sihora, Madhya Pradesh, India

Kamini Lamba
Chitkara University Institute of
  Engineering and Technology
Chitkara University
Rajpura, Punjab, India

Lokesh Malviya
School of Computer Science and Engineering
VIT Bhopal University
Sehore, India

Arvind Mewada
Computer Science and Engineering
  Department
Motilal Nehru National Institute of
  Technology Allahabad
Prayagraj, Uttar Pradesh, India

Anuprita Mishra
Department of Electrical Engineering
IES College of Technology
Bhopal, Madhya Pradesh, India

Amar Nayak
Department of Electrical Engineering
IES College of Technology
Bhopal, Madhya Pradesh, India

Anjali Nighoskar
School of Engineering and
  Technology
Jagran Lakecity University
Bhopal, Madhya Pradesh, India

Dasangam Venkat Nikhil
Department of Computer Science
  Engineering
Technocrats Institute of Technology
  Excellence
Bhopal, Madhya Pradesh, India

Ruchi Pandey
Department of Electrical and
  Electronics Engineering
Gyan Ganga Institute of Science and
  Technology
Jabalpur, Madhya Pradesh, India

Shishir Kumar Shandilya
School of Computer Science
  and Engineering
VIT Bhopal University
Sehore, India

Ananta Narayan Shrestha
Computer Science and Engineering
  Department
Motilal National Institute of
  Technology Allahabad
Prayagraj, Uttar Pradesh, India

Siddharth Shukla
Department of Electrical
  and Electronics
Technocrats Institute of Technology
Bhopal, Madhya Pradesh, India

Nagendra Singh
Department of Electrical and
  Electronics Engineering
Trinity College of Engineering &
  Technology
Karimnagar, Telangana, India

Vandana Sondhiya
University Institute of
  Technology R.G.P.V.
Bhopal, Madhya Pradesh, India

Anita Soni
Department of Computer
  Science Engineering
IES College of Technology
Bhopal, Madhya Pradesh, India

Gireesh Gaurav Soni
Shri Govindram Seksaria Instititute of
  Technology and Science
Indore, Madhya Pradesh, India

Buvaneish Sundar
Department of Computer Science
  Engineering
IES College of Technology
Bhopal, Madhya Pradesh, India

Akash Tiwari
Computer Science and Engineering
  Department
Motilal National Institute of
  Technology Allahabad
Prayagraj, Uttar Pradesh, India

Chapter 1

Faults Diagnosis Using AI and ML

Hameed Khan, Kamal Kumar Kushwah,
Pradeep Kumar Jhinge, Gireesh Gaurav Soni,
Ravi Kant Choubey, and Rupesh Kushwah

Contents

DOI: 10.1201/9781003301820-1

1.1 Introduction

There is a growing tendency in the contemporary commercial production sectors toward the requirement for more incredibly available equipment that can operate continuously around the clock. Therefore, any failure, no matter how slight, cannot be tolerated since it may harm the price and the output. Consequently, it is essential to monitor the machine's state exceptionally carefully and to identify the cause of any failures. Since maintenance was given after the machine had a defect and hindered production, the machine fault detection has significantly improved. Afterwards, AI-based technology changed into regular inspection over the succeeding decades until all industries embraced condition-based maintenance [1]. Performing maintenance on machinery before it develops a malfunction is known as preventive maintenance.

Condition-based maintenance is the practice of doing maintenance with information gleaned from goal measurements. The effectiveness of this method is determined by how well the diagnostic strategies are accurate and implemented [2]. If they want to survive in the current cutthroat market, industries need to increase the dependability of their products while simultaneously lowering manufacturing costs. Product reliability is crucial for specific operations, such as those in the petrochemical, nuclear, and aviation sectors, where any failure might result in catastrophic environmental catastrophes.

Depending on trends and data analysis from one or more factors that signal the development of recognized failures or faults, industries have recently switched from employing the condition-based strategy to the maintenance-based method. An efficient machine condition monitoring approach must detect any issue early on, which must also be able to diagnose the fault's kind and location accurately. The ideal equipment health evaluation provided by the condition monitoring approach must be comprehensive, precise, and complete [3].

However, it would also comprise monitoring temperature, oil analysis, vibration measurement, and acoustic emission (AE) analysis in the conventional sense. In acoustic emission analysis, the transmission medium carries the waves from the emission source to the surface. Electronic signals can be detected from mechanical waves with slight displacement or high frequency. Before the AE equipment processes the data, the signal intensity can be boosted by utilizing a preamplifier.

ANN, FLS, SVM, GA, and others have been widely employed in the field of engineering. If they can be enhanced, AI techniques are helpful compared to typical defect diagnostic methods [4]. These methods are not only more effective, but they can also be readily expanded upon and altered. They may be made adaptable by incorporating new data or knowledge.

Based on the AI and machine learning methodologies, an effort has been made in this chapter to examine current advancements in the field of faults detection and diagnosis of the machine. Both these systems and other conventional methods can be mutually integrated [5, 6].

1.2 AI-Based System

The system that functions like a human being is known as artificial intelligence (AI). It may mimic human behavior as well. It primarily focuses on improving a computer's capacity for cognitive activities, including learning, reasoning, and self-correction. The demand for AI to address engineering issues has grown during the past ten years. These issues were once thought to need human intellect and complicated analytical or mathematical modeling solutions [7]. Modern times have seen a rise in the demand for sophisticated AE analysis tools.

1.2.1 Using Artificial Neural Networks to Diagnose Faults

An approach to information processing is an artificial neural network (ANN). It functions similarly to how the brain processes information in the human body via natural nerve networks. The talk was restricted to introducing the many parts of implementing the ANN. The challenge at hand determined the importance of the network design or topology for ANN performance. Most of the time, choosing the appropriate topology was done using a heuristic model.

The size of the input and output spaces typically hinted at how many nodes were present in the input and output layer. It was crucial to choose between network complexity and regularization. Several parameters need to be selected while creating a neural network. These variables include the number of training iterations, the number of layers, the learning rate, the number of neurons in each layer, the transfer functions, and so on [8, 9].

1.2.2 The Architecture of a Neural Network

After going through a learning phase, ANN has the advantage of being able to react to an input pattern desirably. According to earlier studies, the effectiveness of ANN can anticipate the flaws of machining operations. ANN has been widely used to diagnose mechanical gear, bearings, and rotating machinery health. It relies more on vibration signals' characteristics than valuable data. Advanced AI data analysis tools that can differentiate between diverse AE data sources are in more demand. As a result, cutting-edge unsupervised pattern recognition (UPR) and supervised pattern recognition (SPR) analysis have been combined with classic, graphical AE analysis to create new, more flexible pattern recognition software. The ability to distinguish between noise and damage progression has also improved due to the application of UPR approaches to AE data during various test situations. The issue of a roller with health monitoring has served as an example of how well GA works for fault classification when combined with ANNs. The fault diagnosis systems use acoustic emission and vibration signals as input signals. A prediction method for rotating machinery failure was also used without ANN [10]. The information was shown as signals of processed acoustic emission and vibration. Previous studies examined the use of acoustic emission for the early diagnosis of issues with the dynamic components of the helicopter rotor head. They analyzed the flight test data set's stress wave using wavelet-based methods to compare the results of machinery failure to operational background noise. A novel method of AE source localization was presented to address the problems of velocity and temporal discrepancies. This method was applied to estimate the AE source coordination using the ANN approach to recover signal properties. Vibration and structure-borne stress wave monitoring were done (AE). Principal component analysis (PCA) was used to extract the acoustic emission (AE) signal characteristics. It was also possible to distinguish between the three-valve situations using a feed-forward neural classifier [11].

The AE data acquired during a static test of a 12-m FRP wind turbine blade was analyzed and categorized using several UPR techniques. Based on the UPR findings, SPR algorithm was built using a back-propagation neural network. After the AE data collection, the identical blade underwent a further biaxial fatigue loading. The neural network's learning, interpolation, pattern recognition, and classification capabilities have drawn interest in grinding research. They used neural networks and an aluminum oxide grinding wheel to conduct grinding tests on a surface-grinding machine to classify the machine's burn degrees [12]. The neural networks' inputs consisted of the AE, power data, and statistics from their digital signal processing.

Additionally, the ANN technique was suggested to identify workpiece "burn," an unwelcome modification of the material's metallurgical characteristics brought on by too forceful or another incorrect grinding. The acoustic emission and cutting power were also collected utilizing a quicker sample rate data-gathering device to address the drawbacks of standard ways for monitoring and troubleshooting an unattended milling machine [13]. Data from forces, spindle currents, and acoustic emissions that have undergone some signal processing to determine the membership functions of fuzzy relations were then employed as inputs to neural networks. Fuzzy logic techniques were also used to diagnose the system's condition concerning tool wear and chatter. The results showed that using neural networks for surface roughness prediction and detecting and categorizing workpiece "burn" was promising. The issue of impact damage hurts the composites sector. Although this damage may appear minor, it frequently harms the performance of the composite construction. The impact damage can be located or identified by shape using traditional nondestructive evaluation (NDE) techniques, but its implications on the structure cannot be determined. In contrast, AE captures the active defect development as soon as the structure is loaded.

A novel quantitative analysis concept for pressure vessel AE sources was developed utilizing artificial neural network classification, combined with a fresh approach to determining the severity of the AE sources [14]. The AE signals were recorded via data acquisition, which was then used to weld aluminum alloy plates that were 3 mm thick. The multilayer feed-forward ANN also utilizes wavelet transformations (WT) for the statistical and temporal aspects of the breakdown of EA signals. The detection of partial discharges (PDs), signal processing, and pattern recognition was also investigated using AE measurements and the back-propagation (BP) ANN. Three-dimensional patterns and brief Fourier transformations were used to process the observed signals. The results demonstrated that using BP ANN with the superficial digital flexor tendon components to categorize the various PD patterns produced outstanding results. Several experiments also carried out performance evaluations for the ANN classifier and feature extraction. They were beneficial for data entry during AE studies at the chemical processing facility. There were many AE power spectra included in this input data. Preprocessing was performed on each source input data file, which included extra linear averaging in each input vector and individual amplitude normalization by eliminating the mean value and dividing by the standard deviation of the feature. Back-propagation updating was employed to evaluate the combined feature extraction and classification capabilities of three-layer networks while addressing the issue of process stage recognition [15].

To anticipate the lubricant regime, an artificial neural network and regression models were employed as inputs, together with information on the oil temperature, acoustic emission signals, and a particular film thickness. AE and temperature data were used as input to feed forward back propagation (FFBP) and Elman network models to estimate a certain oil film thickness. The findings demonstrated that the FFBP and Elman models could accurately estimate oil layer thickness from acoustic emission signals and temperature. The recommended method achieved a high prediction and classification success rate of 99.9% during training. During testing, the FFBP outperformed Elman and produced top-notch predictions and classifications. As a result, the design and topology of the network through particular systems may be utilized to forecast any reasons for spur gear operation failure and monitor the oil film thickness in real-time.

1.2.3 Spiking Neural Network

Many academics have lately shown an interest in spiking neural networks (SNNs), third-generation neural networks. The SNNs became well-known before the sigmoid or perceptron neuron was created. SNNs were particularly well-suited for parallel implementation in both digital and analogue hardware.

Older neural network generations employed analogue signals to transmit data from one neuron to the next. The SNNs employed a mechanism similar to human neurons called spikes for inter-neuron communication. The principal neuron employed the weighted sum of the analogue input value to assess the value using a sum-specific non-linear function. The quantity was used to calculate how long the spike output intended for the next neuron would delay. Since the target neuron integrated the spikes for a while and identified the resultant mixed values as the membrane potential, the spiking neuron was frequently referred to as the leaky integrator. The neuron was shown to send a spike when the membrane potential value got close to a particular threshold value; the membrane potential value was then reset. Many other criteria that had to be considered for the neurons to spike could now be explained, thanks to advances in our understanding of how biological neurons interpret information. The other variables included the various connections' physical characteristics, the probability that the spikes would be processed at the synapse, as well as the neurotransmitters that have been produced or the open ion channels. To examine the biological neural system, a number of the attributes mathematically simulated the artificial neurons used in SNNs to communicate used trains, regarded as pulse-coded data. The SNN was expected to give a mechanism to characterize the frequency, time, phase, and other qualities for processing information. It was also thought to be physiologically acceptable. The SNN may also instruct neurons to produce spikes using their spatial-temporal data. The computational effectiveness and the biological plausibility must be considered while choosing the neural model for an SNN. The leaky integrate-and-fire (LIF) model would have to be utilized since it would be more practical if computing efficiency outperformed biological plausibility. The SNN technique was used to demonstrate how the prototype decision support system may track tool wear. The six elements of this system were data gathering, feature extraction, multi-sensor integration, pattern recognition, tool wear evaluation, and outlier detection. They proposed a self-organizing neural architecture based on the SNN with a single integrated component [16]. The modeling technique was quite effective at describing the degree of tool wear on the tool inserts. Their technique showed how useful the SNN model is for monitoring tool performance, proving that the tactic applies to various industrial applications where a lot of noisy data is produced. The results demonstrated the potential of spiking neural networks for fault diagnosis because this researcher was the only one to employ SNN.

1.2.4 Diagnostics of Faults Using Genetic Algorithms

An evolutionary algorithm is a subset of artificial intelligence. Based on a natural selection process that parallels biological evolution, a genetic algorithm may solve optimization problems that are both confined and unconstrained. An individual population of solutions is modified periodically by the algorithm. The genetic algorithm (GA) chooses individuals randomly from the existing population at each stage, using them as parents to create offspring for the following generation. People "evolve" toward the best option over future generations.

As initially proposed, the three major processes that make up a simple GA are replacement, genetic operation, and selection. The collection of chromosomes that made up the population was the answer's contender. All chromosomes' fitness values were assessed in a decoded form by an objective function. Based on the known genetic processes of crossover and mutation, a particular set of parents was chosen from the population to produce children.

1.3 Genetic Algorithm Cycle

The chromosomes in the current population were then replaced by their offspring using a specified replacement technique. Until the termination condition was met, this GA cycle was repeated. To demonstrate the efficiency of GA in AE feature selection for fault classification, ANNs were employed to depict a straightforward scenario, including a roller with health monitoring. Using Gas was the most effective strategy for deciding the suitable feature set for an ANN classification application. For bearing condition monitoring and problem diagnosis, Ming uses the AE approach, which uses the wavelet-based waveform parameter selection, continuous wavelet transform scales, and genetic algorithm-based optimization. While performing the mechanical tests on various materials, the AE was monitored by using a data-gathering system. Two of the sensors were put on the specimen itself. After AE signals were captured during the experiments, "model" data sets were created. A genetic algorithm-based method was described and verified to cluster the AE signals. The analysis of several "model" data sets demonstrated its superiority over the k-means technique. The evolutionary approach is efficient and stable at clustering data sets that include members of a minority class, a cluster with extreme feature signals, or a collection of groups with vastly different sizes.

1.3.1 Fault Diagnosis Based On Fuzzy Logic

A multi-valued logic that allows for values between the typical true/false, yes/no, high/low, and other assessments is called fuzzy logic (FL). The FL offers several solutions to control or categorize problems. Therefore, rather than attempting to simulate how the system functions, this approach focuses on what it should accomplish. The surface grinding machine's burn degrees were categorized using artificial neural networks (ANNs).

An adaptive neuro-fuzzy inference system was used to develop a method for predicting the surface roughness of advanced ceramics. Rectangular bars made from alumina workpieces were crushed and sintered for this investigation [17]. Additionally, it uses the statistical data derived from the AE signal and cutting power as input. They offered strategies for a one-board fault detection system and a test program set (TPS) fault detection system for electromechanical actuator (EMA) ball bearings by analyzing the various vibration and AE data and utilizing FL inference techniques [18].

They demonstrated the results of fuzzy modeling to pinpoint the grinding problem by digital processing of the produced acoustic emission signals. The AE signal was divided into many signal sources using fuzzy C-means (FCM) clustering. When the borders of the subgroups overlap, FCM may have helped identify the cluster in the data. The surface of a solid steel block was used for the AE test, which was performed using pulse, pencil, and spark signal sources. The AET 5000 system measured four characteristics: event length, peak amplitude, rise time, and ring down count. The FCM-based classification was then trained on data and validated.

Based on a fuzzy model, the raw AE and cutting power data were digitally processed to provide the models' inputs. The mean-value deviation, grinding power, and root mean square (RMS) of the acoustic emission signal were the characteristics of fault gathered. They attempted to use the best AE model throughout the continuous cutting periods by using fuzzy modeling. The fuzzy identification technique provided an easy way to draw a more definitive conclusion from the collected data, given the challenges of knowing the precise physics of the machining process. Because type-2 fuzzy logic requires exceedingly fuzzy conditions, recent studies have employed when type-1 fuzzy methods cannot sufficiently model in the field of research. Type-2 FL would be used if we were to employ FL at a higher level. To filter the raw AE data straight from the AE sensor while turning, they adopted the type-2 TSK (Takagi–Sugeno–Kang (TSK)) fuzzy uncertainty estimates approach. The filtering and capture of uncertainty by type-2 TSK fuzzy technique on the interval of AE signal throughout one 10 mm cutting length is the specific emphasis. They demonstrated a type-2 FL application to modeling AE signals in precision production. For differentiating the AE signal in precision machining, type-2 fuzzy modeling was applied. Without knowing the precise mechanics of the machining process, it offered a straightforward method for reaching a firm conclusion. Uncertain tool life forecast information was essential for examining tool conditions. It was also making judgments on how to preserve the machine's quality [19].

To accurately predict the cutting tool condition throughout its life, type-2 FLSs were used in the system to analyze the AE signal feature (SF) and select the most trustworthy ones for integration. According to the obtained results, the type-2 fuzzy tool life estimation is consistent with the level of cutting tool wear during the micro-milling operation.

The AE SFs in TCM were analyzed using a type-2 fuzzy analytic approach during the micro-milling process. The type-2 approach's interval output supplied an interval of uncertainty related to SFs of the AE signal. To predict the cutting tool life in the future, the SFs with the lowest root mean square error and variation was chosen. A new approach to localizing AE sources in environments with significant background noise was also developed. The method was built on fuzzy-neuro and probabilistic principles, allowing AE events to be categorized according to their energy and location likelihood. For testing novel algorithms, AE signals captured during the stamping of a thin metal sheet were employed.

An efficient walnut recognition system was developed by combining the AE analysis, principal component analysis (PCA), and adaptive neuro-fuzzy inference system (ANFIS) classifier. Selected statistical characteristics were fed into the ANFIS classifier during the classification phase.

1.3.2 Using Support Vector Machines to Diagnose Faults

Based on the statistical learning theory, the support vector machine (SVM) method was used as a classification method. It was primarily based on the theory of linear hyperplane classifier. SVM's primary goal was to investigate a linear ideal hyperplane for increasing the distance between the two classes. The spur bevel gear box's issue was diagnosed using the SVM. Due to its greater accuracy and generalization skills, this was regarded as a prominent machine learning application. This research also looked at low-speed bearing defect diagnostics using the AE method and vibration signal. The classification method was used to undertake fault diagnosis using relevance vector machines (RVMs) and SVMs. To diagnose low-speed bearing faults, the classification procedure allowed for a comparison between RVM and SVM [20].

The classifier was created to identify and categorize AE signals. The outcomes of the simulation demonstrated that SVM might potentially discriminate between various acoustic emission signals and noise signals efficiently. Using this approach, grid search parameters had a greater classification accuracy rate than the GA algorithm. Based on a thorough review of the literature, by creating fractures in rock samples during a surface instability test, this approach provides novel techniques for grouping AE signals and identifying P-waves for the location of microcracks in the presence of noise. These methods are based on hierarchical clustering and SVMs. In light of this, the suggestions are unique discrete wavelet decomposition and SVM-based grinding wheel wear monitoring method. An AE sensor was used to gather the grinding signals.

The development of the AI approach indicates excellent potential in machine status monitoring and diagnostics. However, several pertinent issues have effectively used ANN based on AE. It may be said that ANN is the most recent and well-liked technique for AE signal condition monitoring [21].

1.3.3 Concept for Remote Fault Monitoring and Detection

Early fault detection is essential to keeping the system operational for a long time since some defects can lead to system failure when they occur repeatedly. System modeling and model assessment are the foundations of model-based approaches. In approaches based on signal processing, defect information is extracted from relevant signal properties using mathematical, statistical, or artificial intelligence techniques. Since sensor data may be transferred to the processing center via various techniques and provides in-situ measurements, feature-based approaches are best suited for remote monitoring.

Information defining each monitored component's condition must create a trustworthy fault-detection process utilizing feature-based approaches. The information is derived from numerous sensor signals. Acoustic emission, torque, strain, electrical output, and lubricating oil quality may be employed.

1.3.4 Diagnostic of Faults and Process Flow

There are following main steps involved:

- Access and preprocess data: Preprocessing is necessary since not all sensor data received from equipment is typically helpful.
- Synchronization of time series data: aligning data that may include missing values or data gathered at various speeds.
- Advanced sensor data noise removal.
- Extracting, transforming, and choosing features: to select the information that will be most helpful in predicting failure.
- Create fault detection models.
- Fault detection models are built using data clustering, classification, and system identification techniques based on mathematics, statistics, or artificial intelligence. Predictors and response data are used to train, verify, and test these models.
- Models are then deployed in the production environment, with the option of focusing on real-time embedded hardware to corporate systems, PCs, clusters, and clouds.

1.4 AI Model Development for Fault Detection

Artificial intelligence methods like machine learning and neural networks may find faults using sensor data. These strategies allow for implicit programming-free learning utilizing training data. The newly acquired sensor data may then produce predictions using the trained algorithm [22].

Machine learning tasks are primarily divided into two groups:

1. **Supervised learning:** The algorithm learns a general rule that maps inputs to outputs by being trained on examples of inputs and the desired outputs. Supervised learning for fault detection technique is shown in Figure 1.1.
2. **Unsupervised learning:** In this case, the learning algorithm makes its own categorization decisions to detect input structures because labels are not given to it. Unsupervised learning is mainly employed to find data's hidden patterns.

Supervised learning is the most suitable method in wind turbines since it allows for training fault detection algorithms using historical sensors and associated fault occurrence data as predictors.

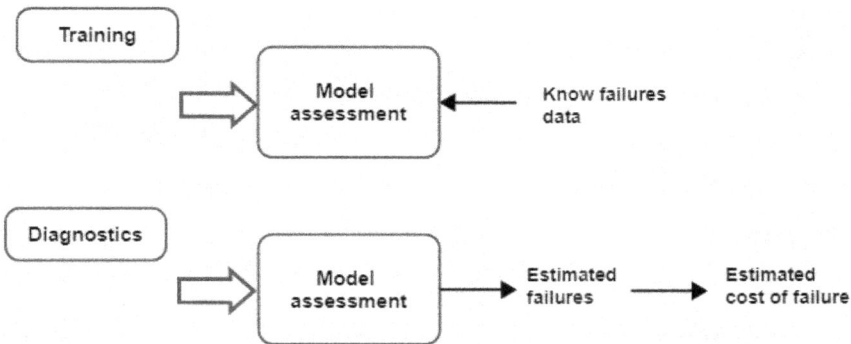

Figure 1.1   Supervised Learning for Fault Detection.

There are two significant groups of algorithms for supervised learning:

- **Classification:** for responses with categorical values, where the information may be divided into distinct "classes."
- **Regression:** for predicting continuous response values.
- Since the sensor data is utilized to forecast a definite answer, "fault" or "no-fault," the fault detection problem comes within the classification category. Different categorization algorithms are readily available. However, the following two methods are the most effective at using sensor data to identify faults in wind turbines: artificial neural networks (ANNs) and support vector machines (SVMs).

1.4.1 Support Vector Machines (SVMs)

An SVM is used to categorize the data, and the best hyperplane that separates all the data points of one class from the other is discovered. The ideal hyperplane for an SVM is the one with the most significant margin between the two classes. We define the margin as the largest width of the slab perpendicular to the hyperplane without any inside data points. The data points nearer to the separating hyperplane are known as support vectors. These data points are located on the slab's edge. These definitions are shown in Figure 1.2.

A linear classifier or a classifier that shown in Figure 1.3 uses a line to divide a collection of items into their respective categories. The majority of classification jobs, however, require more complicated and frequently more nonlinear structures to get the best separation.

Compared to the previous schematic, a curve will be needed to fully separate the items. Support vector machines carry out the separation process in this scenario.

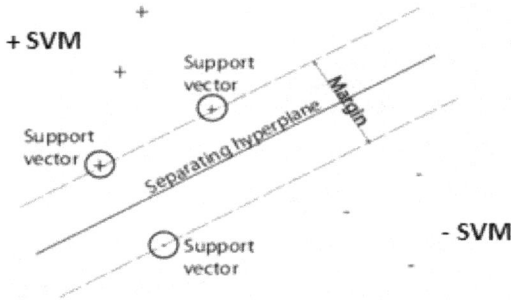

Figure 1.2   SVM Concept.

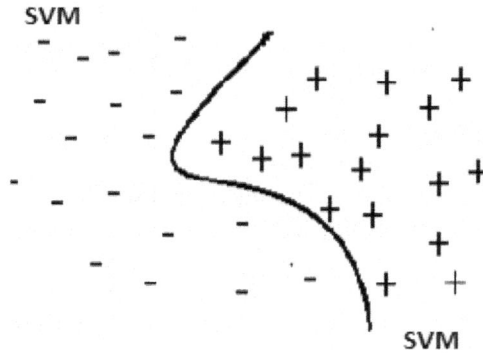

Figure 1.3   Non-linear Separation in SVM.

As shown in Figure 1.4, a collection of mathematical operations known as kernels is used to map, that is, rearrange or organize the initial items depicted on the left side of the graphic. In this new scenario, the mapped items are linearly separable. This is why, instead of creating the intricate curve, it is essential to locate an ideal line that may divide the items (left schematic) [23].

1.4.2 *Challenges*

1.4.2.1 *The Ability to Generalize*

As was already established, it is challenging to precisely forecast the remaining useful life (RUL) of a new item due to the few-shot data. The capacity of the RUL estimate to generalize has been improved by the application of transfer learning, although mainly from the feature or model perspective. Increasing the amount of data is a logical strategy to improve this capability. It is still challenging to produce

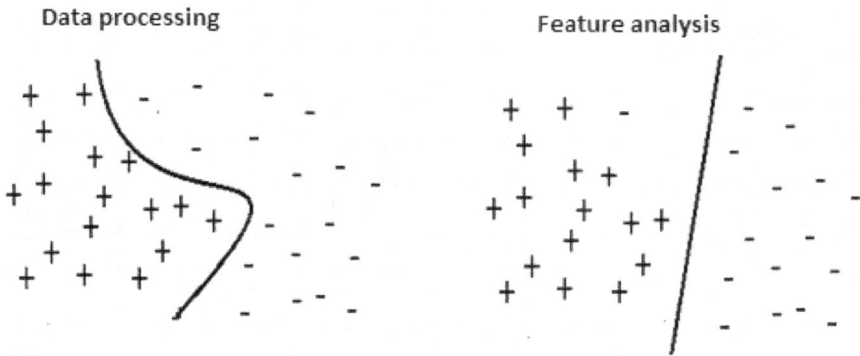

Figure 1.4 Transition from Nonlinear to Linear.

more high-quality samples using the available information. Although it is hard to collect the exact data distribution, there are ways to estimate it, such as by creating high-quality virtual samples using resampling or machine learning strategies. Additionally, various degradation processes under multiple situations may be developed by altering the operating conditions, fault kinds, and other factors with the digital twin model of a particular mechanical system.

1.4.2.2 Prognosis in Real-World Scenarios

The current open world has various constraints and uncertainties, such as limited computer resources, unstable working circumstances, unidentified failure nodes, etc. Due to a lack of processing capability, AI-enabled forecasting techniques cannot be directly applied to the current scenario. Meanwhile, a lightweight model must be created to ensure real-time prediction; typical methods include model compression and pruning. Additionally, since the open design and training data distribution is frequently uneven, the model must be able to update parameters online continually.

1.4.2.3 Combining Model-Driven and Data-Driven Approaches

As equipment complexity rises, it is often difficult to analyze and anticipate the RUL using a single technique. We can fully exploit the strong feature extraction capabilities based on data-driven strategies and the benefits of interpretability of model-driven methods to construct more beneficial health indicators by merging different models based on data-driven and model-driven approaches [24]. To predict wind turbine primary bearing fatigue, it has developed a physics-informed layer based on damage growth in deep neural networks and released a cross-physics data fusion system and a loss function.

1.5 Conclusion

The machine learning techniques have played a very important role for automated fault detection and diagnosis in industrial machinery under real-world conditions. When conducting problem detection and prognosis in the manufacturing industry, it is imperative to consider the inherent complexity of manufacturing systems and the time-varying nature of production processes. Algorithms and machine learning techniques can effectively handle noisy, non-stationary data and record the non-linear patterns of interaction between machinery parts. It is recommended to use online learning algorithms that learn gradually or from small batches of recent data to process high-speed streams of sensor data while continuously adapting to changes in the data's probability distribution caused by non-stationary conditions. Concept drift will eventually make learning models obsolete if they are not considered. In the chapter, the authors presented that there is currently a shortage of research on online learning techniques, particularly regarding the identification or prediction of mechanical problems in the manufacturing sector. As a result, this field of study offers exciting potential for further investigation.

The analysis of machine learning (ML) and artificial intelligence (AI) applications in fault diagnosis showed that ML approaches may be a very helpful tool for fault identification and diagnosis. This study demonstrates that not all the features needed to diagnose a system can be filled by a single approach. To improve the diagnostic system, certain approaches can supplement others. With the Internet of things, sensors produce a vast amount, diversity, and velocity of data in the era of industry 4.0. Additionally, a cloud must be planned for data storage. But there are certain issues, such as scalability and security. Therefore, it's crucial to create a diagnosis system, which can spot and pinpoint the defect caused by a system failure. The Advanced AI and ML tools can handle this difficulty and provide a thorough diagnostic in the future.

References

1. Seongmin Heo, & Jay H. Lee (2018) Fault Detection and Classification Using Artificial Neural Networks. IFAC-PapersOnLine. 51(18): 470–475. ISSN 2405–8963. https://doi.org/10.1016/j.ifacol.2018.09.380.
2. S.R. Prasad (March 3, 2022) Vishnu and Poulose, Jasmine and Sadique, Anwar, Fault Detection of Spur Gear Using Machine Learning. http://dx.doi.org/10.2139/ssrn.4048601.
3. G. Hu, T. Zhou, & Q. Liu (2021) Data-Driven Machine Learning for Fault Detection and Diagnosis in Nuclear Power Plants: A Review. Front. Energy Res. 9: 663296. Doi: 10.3389/fenrg.2021.663296.
4. R. Argawal, D. Kalel, M. Harshit, A.D. Domnic, & R.R. Singh (2021) Sensor Fault Detection Using Machine Learning Technique for Automobile Drive Applications. NPEC: 1–6. Doi: 10.1109/NPEC52100.2021.9672546.

5. A.A. Silva, A.M. Bazzi, & S. Gupta (2013) Fault Diagnosis in Electric Drives Using Machine Learning Approaches. 2013 IEMDC: 722–726. Doi: 10.1109/ IEMDC.2013.6556173.

6. Prashant Kumar, & Ananda Hati (2020). Review on Machine Learning Algorithm Based Fault Detection in Induction Motors. Arch. Comput. Methods Eng. 28. Doi: 10.1007/s11831-020-09446-w.

7. David Verstraete, Andrés Ferrada, Enrique López Droguett, Viviana Meruane, & Mohammad Modarres (2017) Deep Learning Enabled Fault Diagnosis Using Time-Frequency Image Analysis of Rolling Element Bearings. Shock Vib. 2017: 17. Article ID 5067651. https://doi.org/10.1155/2017/5067651.

8. M. Fernandes, J.M. Corchado, & G. Marreiros (2022) Machine Learning Techniques Applied to Mechanical Fault Diagnosis and Fault Prognosis in the Context of Real Industrial Manufacturing Use-Cases: A Systematic Literature Review. Appl. Intell. https://doi.org/10.1007/s10489-022-03344-3.

9. M.S. Kan, A.C.C. Tan, & J. Mathew (2015) A Review on Prognostic Techniques for Non-Stationary and Non-Linear Rotating Systems. Mech. Syst. Signal Process. 62: 1–20. https://doi.org/10.1016/j.ymssp.2015.02.016.

10. H. Hu, B. Tang, X. Gong, W. Wei, & H. Wang (2017) Intelligent Fault Diagnosis of the High-Speed Train with Big Data Based on Deep Neural Networks. IEEE Trans. Ind. Inf. 13(4): 2106–2116. https://doi.org/10.1109/TII.2017.2683528.

11. H. Shao, H. Jiang, F. Wang, & H. Zhao (2017) An Enhancement Deep Feature Fusion Method for Rotating Machinery Fault Diagnosis. Knowl-Based Syst. 119: 200–220. https://doi.org/10.1016/j.knosys.2016.12.012.

12. G.A. Susto, A. Schirru, S. Pampuri, S. McLoone, & A. Beghi (2015) Machine Learning for Predictive Maintenance: A Multiple Classifier Approach. IEEE Trans. Ind. Inf. 11(3): 812–820. https://doi.org/10.1109/TII.2014.2349359.

13. A. Krishnakumari, A. Elayaperumal, M. Saravanan, & C. Arvindan (2017) Fault Diagnostics of Spur Gear Using Decision Tree and Fuzzy Classifier. Int. J. Adv. Manuf. Technol. 89(9–12): 3487–3494. https://doi.org/10.1007/s00170-016-9307-8.

14. Z. Li, Y. Wang, & K-S. Wang (2017) Intelligent Predictive Maintenance for Fault Diagnosis and Prognosis in Machine Centers: Industry 4.0 Scenario. Adv. Manuf. 5(4, SI): 377–387. https://doi.org/10.1007/s40436-017-0203-8.

15. Leo H. Chiang, Mark E. Kotanchek, & Arthur K. Kordon (2004) Fault Diagnosis Based on Fisher Discriminant Analysis and Support Vector Machines. Comput. Chem. Eng. 28(8): 1389–1401. ISSN 0098–1354. https://doi.org/10.1016/j. compchemeng.2003.10.002.

16. Ahmed Ragab, Mohamed El Koujok, Hakim Ghezzaz, & Mouloud Amazouz (2021) Chapter 10—Fault Diagnosis in Industrial Processes Based on Predictive and Descriptive Machine Learning Methods, Editor(s): Jingzheng Ren, Weifeng Shen, Yi Man, & Lichun Dong, Applications of Artificial Intelligence in Process Systems Engineering, Elsevier, Pages 207–254. ISBN 9780128210925. https://doi.org/10.1016/ B978-0-12-821092-5.00002-4.

17. Vinny Foba, Alexandre Boum, Camille Mbey, et al. (June 29, 2022) Fault Detection and Classification Using Deep Learning Method and Neuro-Fuzzy Algorithm in a Smart Distribution Grid. Authorea. Doi: 10.22541/au.165649366.60373963/v1.

18. S. Manikandan, & K. Duraivelu (2021) Fault Diagnosis of Various Rotating Equipment Using Machine Learning Approaches—A Review. Proc. Inst. Mech. Eng. Part E: J. Process Mech. Eng. 235(2): 629–642. https://doi.org/10.1177/0954408920971976.

19. B. Luo, H. Wang, H. Liu, B. Li, & F. Peng (2019). Early Fault Detection of Machine Tools Based on Deep Learning and Dynamic Identification. IEEE Trans. Ind. Electron. 66: 509–518.
20. Daudi Mnyanghwalo, Herald Kundaeli, Ellen Kalinga, & Ndyetabura Hamisi James Lam (Reviewing editor) (2020) Deep Learning Approaches for Fault Detection and Classifications in the Electrical Secondary Distribution Network: Methods Comparison and Recurrent Neural Network Accuracy Comparison. Cogent Eng. 7: 1. Doi: 10.1080/23311916.2020.1857500.
21. M. He, D. He, & E. Bechhoefer (2016) Using Deep Learning Based Approaches for Bearing Fault Diagnosis with AE Sensors. Annual Conf. PHM Soc. 8(1). https://doi.org/10.36001/phmconf.2016.v8i1.2569.
22. M. Jamil, S.K. Sharma, & R. Singh (2015) Fault Detection and Classification in Electrical Power Transmission System Using Artificial Neural Network. SpringerPlus. 4: 334. https://doi.org/10.1186/s40064-015-1080-x.
23. V. Handikherkar, & V. Phalle (2021). Gear Fault Detection Using Machine Learning Techniques- A Simulation-Driven Approach. Int. J. Eng. 34(1): 212–223. Doi: 10.5829/ije.2021.34.01a.24.
24. W. Zhang, D. Yang, & H. Wang (2019) Data-Driven Methods for Predictive Maintenance of Industrial Equipment: A Survey. IEEE Syst. J. 13(3): 2213–2227. https://doi.org/10.1109/JSYST.2019.2905565.

Chapter 2

Load Frequency Control for Multi-Area Power System Using PSO-Based Technique

Siddharth Shukla, Amit Gupta, and Ruchi Pandey

Contents

DOI: 10.1201/9781003301820-2

2.1 Introduction

The main aim of using load frequency control (LFC) is to ensure a sensible uniform frequency while balancing the load among the generators and controlling the tie-line power to prespecified values. LFC in the power system is essential for providing reliable and consistent electric power. The distinguishing characteristic of a general operating system is constant frequency.

A power plant has to keep count of the load conditions and provide customers with service all day. The idea that power is generated uniformly everywhere is therefore irrelevant. Power generation thus changes depending on the load. The goal of a control strategy is to create and deliver power in a networked system as efficiently and dependably as possible, while keeping the frequency and voltage within acceptable ranges. While reactive power depends on variations in voltage magnitude and is less sensitive to frequency, it is affected by changes in load most significantly. To maintain the same frequency, the turbines used to tune the generators are controlled by a proportional plus integral (P-I) controller, and the steady state inaccuracy of the system frequency is also decreased by adjusting the controller gains.

There are various algorithms, such as genetic algorithm (GA), which optimize the controller gains for load frequency management of an interconnected power system, but GA is challenging to implement due to its challenging coding and slow convergence rate. Bacterial foraging optimization algorithm (BFOA) is a different technique for solving the reproduction problem, which results in a population of N individuals. Because it is simple, does not have an impact on the magnitude of the problem, and effectively solves large-scale non-linear optimization problems, particle swarm optimization (PSO) is employed in this work.

Since 50 Hertz is the normal operating frequency in India, each deviation of 2.5 Hertz will have a significant impact on the entire system. Turbine blades, for instance, are vulnerable to damage in such circumstances. Additionally, there is a relationship between frequency and motor speed, and frequency change will also have an impact on this relationship. The objective of this work is

■ To create a controller based on the PSO algorithm's optimum parameters for reducing the frequency value to a fixed value against any variation in load demand
■ The power flow through the tie-line of each area must be maintained to its prespecified value
■ Minimize the error of the system

2.2 Literature Survey

The primary goal of power system operation and control is to retain power of an acceptable quality. One of the most crucial issues in the connected power systems is the load frequency control (LFC) issue. According to the literature, C. Concordia and L.K. Kirchmayer et al. [1] have put a lot of effort on studying LFC. Olle I. Elgerd does excellent work on LFC of power systems [2].

The power system is a complicated, nonlinear system that experiences a variety of occurrences. For a stable power supply despite variations in load, a power system's frequency must be maintained constant. Numerous different control algorithms have recently been put out for LFC. The genetic algorithm [3, 4] is a reliable and adaptive method for solving search and optimization problems; however, because of its complexity in coding, both its implementation and convergence time are challenging.

By adjusting the gains of the controller, the bacterial foraging optimization algorithm is another method for keeping the frequency within acceptable bounds [5]. Since successful foraging strategies are more likely to produce success, the selection process in this process tends to favor the transfer of the genes of animals with successful gathering strategies. Because it involves reproduction, it generates a population of N individuals, which increases the number of parameters that must be set, which is a downside. This brings us to particle swarm optimization (PSO), a population-based approach that was first introduced by James Kennedy and Russell C. Eberhart (1995). This article [6] integrates social interaction theory with problem-solving. PSO is typically chosen because it is simple in concept, straightforward to implement, and computationally efficient. Thus, PSO is successfully used to adjust the controller's parameters, aiding in the achievement of the goal of maintaining frequency stability by reducing the objective function. Here an interconnected power system's load frequency control (LFC) is implemented using a proportional-integral controller [7]. Increasing proportional gain can reduce offset, but it may also cause oscillations to become more pronounced. In contrast, if the integral control order rises, instability may result. A performance metric used to develop control systems is the Integral of Time multiplied by Absolute Error (ITAE). Graham and Lathrop (1953) developed a set of normalized transfer function coefficients to reduce the ITAE requirement for a step input, which led to the creation of the index [8].

2.3 Load Frequency Control

Electricity is generated via power systems using renewable or natural energy. To provide dependable and superior electric power at the consumer end, load frequency control, or LFC, is essential in power systems. However, the loads are varied randomly and regularly by the electricity consumers. A problem is controlling the power generation, but a load change causes generation to be adjusted so that there is no power imbalance. A control system is necessary to mitigate the impact of haphazard load variations and maintain the voltage and frequency within predetermined ranges. Voltage is related to reactive power, but frequency is closely tied to the real power balance. Load frequency control is the name given to the actual power and frequency management [1]. If the load in a system changes, the frequency and bus voltages will also change as a result. LFC, as its name implies, maintains a constant frequency while adjusting the power flow between various locations. The LFC is a loop that controls the generator's frequency and output in the megawatt range [9]. There are two loops in this—the major loop and the secondary loop. More significant than those of single-area systems are the issues with frequency control of connected areas. Reasons to hold frequency constant are as follows [15]:

1. Most types of ac motors run at speeds that are related to the frequency directly.
2. If normal frequency is 50 Hertz and the turbine runs at speeds corresponding to ±2.5 Hertz, then the blades of the turbine are likely to get damaged.

The synchronous motors power the electrically powered clocks. These clocks' accuracy depends on their frequency as well as a component of their frequency error. Power systems today are interconnected with nearby places. However, the integration of the electricity systems increases the system's order tremendously. Tie-lines make this connection possible.

Tie-lines allow the transfer of power between locations as shown in Figure 2.1. Tie-line power exchange mistake is a problem brought on by the introduction of tie-line power. An area will get electricity from tie-lines from other places when there is a change in load in that area. The power flow across various tie-lines is planned or predetermined, so area i may provide area j with a certain amount of power while removing a different specified quantity from area k. LFC must therefore also manage the tie-line power exchange error. The tie-line power fluctuations are considered to include information about the immediate surroundings. As a result, in a two-area system, the tie-line power is observed, and the resulting tie-line power is distributed back into both areas. As a result, when modeling such complex high-order power systems, the model and parameter approximations cannot be avoided [2]. Modeling of two area system is shown in Figure 2.2.

Hence, LFC has two main objectives:

1. To keep the frequency constant against any load change.
2. Flow of power in the tie-line must be maintained to its desirable value in each area.

The present control objective is to regulate each area's frequency while also controlling the tie-line power in accordance with interarea power contracts. Control of the turbines that turn the generators is done in the case of frequency control or for

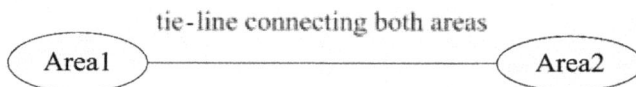

tie-line connecting both areas

Area1 ———————————————— Area2

Figure 2.1 Two Areas Connected by Tie-Line.

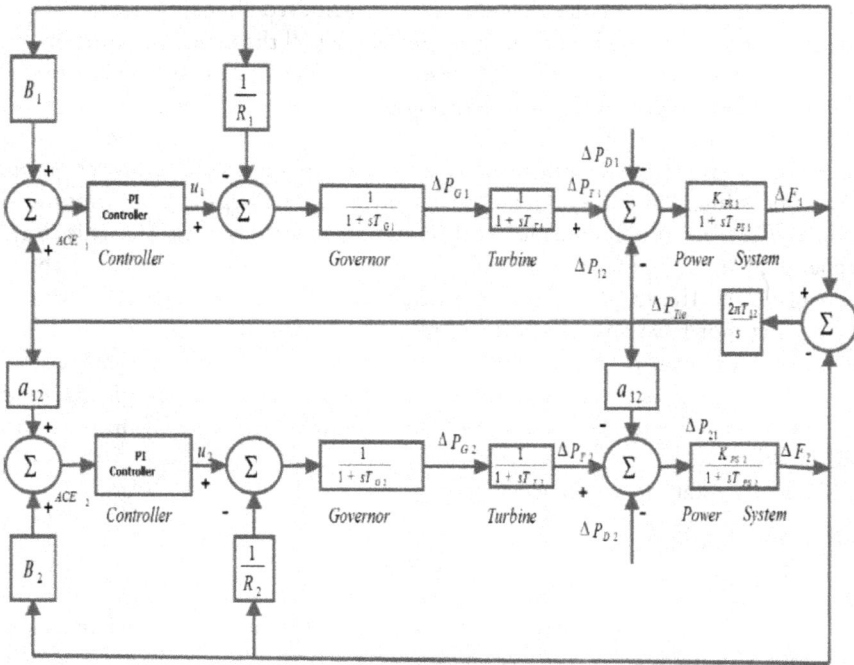

Figure 2.2 Block Diagram of Two-Area Interconnected System.

bringing frequency deviation back to the desired level. To achieve zero steady state error in tie-line power flow, the proportional plus integral controller is commonly employed for this purpose [16].

2.4 Tuning of Controller

Tuning a controller is the process of choosing controller parameters such that it satisfies desired performance requirements. A controller must be tuned for fast response and strong stability. The first tuning guidelines for controllers were put forth by Ziegler–Nichols (Z–N). The fact that the Z–N method entails trial and error, which is undesirable and not suitable to open loop unstable systems, is what spurs the creation of other tuning methods after it. Following that, many fresh methods—such as amigo tuning, analytical tuning, and optimization strategies—were created. Methods based on optimization are becoming more popular today [7].

The traditional approach to controller tuning relies on trial and error. This makes it challenging to execute in a variety of issues. This is a time-consuming procedure since it involves constant adjustment of the parameter values and observation of the reaction. There is no assurance that the outcome achieved after such a wide range of possibilities would be the best one. That implies that all the time and effort invested in a certain issue will be in vain. All these drawbacks push us to find out recently created methods to obtain the best results [17].

Linear quadratic regulator (LQR) is an optimal controller. To achieve optimality, one or more of the plant's outputs must be provided with the least amount of error possible, while the control output must also be kept to a minimum. The model of the plant under consideration or management is the basis for this control mechanism. If the model precisely duplicates the plant, the controller is considered optimum. The LQR is a state feedback controller in which a system's states may or may not have any physical significance. Finding the states to employ for feedback may therefore be difficult. For this another function, called an observer, is required, which estimates the values of the state. But the complexity of the system increases due to involvement of the observer. This is based on state space model [10].

Where, the state space equation is

$$X = Ax + Bu \tag{2.1}$$

and x = state vector u = control vector, control vector "u" is obtained by linearly combining all the states; here, there are nine state vectors [8].

$$u = -Kx \tag{2.2}$$

Where K is the feedback matrix that is to be determined such that the performance index is minimized to fulfill our objective. K is obtained from the solution of a set of linear algebraic equation given by a Riccati equation that is shown below:

$$A^T S + SA - SBR^{-1}B^T S + Q = 0 \tag{2.3}$$

$$K = R^{-1}B^T S \tag{2.4}$$

The value of K for which the system remains stable is acceptable.

$$R = kI \tag{2.5}$$

k is the weighing factor.

Where,
A, B, R, Q, and S are matrices whose dimensions depend on the number of states; R and Q are symmetric matrices.

$$A = \begin{bmatrix} \dfrac{-1}{T_{ps1}} & \dfrac{K_{ps1}}{T_{ps1}} & 0 & 0 & 0 & 0 & \dfrac{-K_{ps1}}{T_{ps1}} & 0 & 0 \\[2mm] 0 & \dfrac{-1}{T_{T1}} & \dfrac{1}{T_{T1}} & 0 & 0 & 0 & 0 & 0 & 0 \\[2mm] \dfrac{-1}{R_1 T_{G1}} & 0 & \dfrac{-1}{T_{G1}} & 0 & 0 & 0 & 0 & 0 & 0 \\[2mm] 0 & 0 & 0 & \dfrac{-1}{T_{pG2}} & \dfrac{K_{ps2}}{T_{pG2}} & 0 & \dfrac{a_{12}K_{ps2}}{T_{pG2}} & 0 & 0 \\[2mm] 0 & 0 & 0 & 0 & \dfrac{-1}{T_{T2}} & \dfrac{1}{T_{T2}} & 0 & 0 & 0 \\[2mm] 0 & 0 & 0 & \dfrac{-1}{R_2 T_{G2}} & 0 & \dfrac{-1}{T_{G2}} & 0 & 0 & 0 \\[2mm] 2\pi T_{12} & 0 & 0 & -2\pi T_{12} & 0 & 0 & 0 & 0 & 0 \\[2mm] b_1 & 0 & 0 & 0 & 0 & 0 & 1 & 0 & 0 \\[2mm] 0 & 0 & 0 & b_2 & 0 & 0 & -a_{12} & 0 & 0 \end{bmatrix}$$

$$B^T = \begin{vmatrix} 0 & 0 & \dfrac{1}{T_{G1}} & 0 & 0 & 0 & 0 & 0 & 0 \\[2mm] 0 & 0 & 0 & 0 & 0 & \dfrac{1}{T_{G2}} & 0 & 0 & 0 \end{vmatrix}$$

One drawback of the LQR approach is that the user must supply a weighted factor to minimize a function. The user's laborious task in controller optimization is to set the weight factor and compare the outcomes. Finding the appropriate weighing factor is a challenge, which is another drawback.

2.5 Optimization Algorithms

The effort and time needed to obtain the controller's optimum values using the aforementioned approaches push us to choose a more sophisticated approach that uses an optimization algorithm based on natural processes. Numerous optimization algorithms have been proposed or implemented to date, including evolutionary programming, genetic algorithms, and heuristic methods. The genetic algorithm (GA), bacterial foraging optimization algorithm (BFOA), and particle swarm optimization (PSO) algorithms are the key recently published techniques used in the study and comparison.

2.5.1 Genetic Algorithm

John Holland and his students first introduced genetic algorithms (gAs). These algorithms essentially use an adaptive strategy to look for and resolve optimization issues. This is based on biological organisms' genetic processes. People in a population compete with one another for food, shelter, and mates. Those who are most effective in both surviving and attracting mates will produce a large number of offspring. Does this mean that each succeeding generation will carry the DNA of healthy persons? Each person receives a fitness score based on how well-suited a solution it offers to a certain situation. This can be applied to tasks like image processing, machine learning, and pattern identification. This is a strong technique that can successfully manage a wide range of issue areas that are challenging for other techniques. Finding an efficient solution to an issue in a short amount of time is effective. This algorithm's performance is influenced by factors including population size, the number of generations, mutation rates, and crossover rates. The likelihood of discovering a global optimum solution increases with large population size and generation, while processing time increases significantly [11].

Advantages:

- This approach is easily adaptable to current simulations and models.
- This approach can resolve all chromosomal encoding-related issues.
- There is no assurance that it will provide a global optimum.
- Requires thorough understanding of fitness function. The problem cannot be addressed if the fitness function is unknown.
- Low convergence speed and coding complexity make it difficult to implement, and a few of its drawbacks include premature convergence and slow completion.

Disadvantages:

- There is no assurance that it will provide a global optimum.
- It requires a thorough understanding of fitness function. The problem cannot be addressed if the fitness function is unknown.
- Implementation is challenging because of the high coding complexity and slow convergence speed.
- Its drawbacks include sluggish finishing and premature convergence.

2.5.2 Bacterial Foraging Optimization Algorithm

Passino suggests the bacterial foraging optimization algorithm (BFOA). Typically, bacteria look for nutrients to optimize their energy. Chemotaxis, swarming, reproduction, and elimination-dispersal are the four primary BFOA processes.

Chemotaxis is the process by which bacteria migrate in search of nutrition, and the core concept of BFOA is to mimic the chemotactic movement of bacteria in the search space. Bacteria can move, tumble, or swim when they are foraging, thanks to flagella. Flagella rotating counterclockwise pulls the cell, causing separate flagella movement. The bacteria then tumble faster to find a nutritional gradient in a risky location. When the flagella move in the opposite direction, the bacterium swims quickly. When there is enough food available, they grow longer and, if the temperature is right, split in half.

2.5.3 Particle Swarm Optimization

Genetic algorithms' premature convergence issue is resolved by BFOA's dispersal and elimination method, which assists in locating acceptable sections when the population involved is small. This BFO algorithm's primary drawback is that it uses a reproduction method that generates a population of N individuals as a result of the numerous parameters that must be set.

The concept of particle swarm optimization (PSO) was developed in 1995 by James Kennedy and R.C. Eberhart. It is an evolutionary computation technique that uses stochastic (connections of random variables) computation to search the search space. This method is based on the mobility and intelligence of swarms. This is a population-based strategy because it is based on swarm behavior. When seeking for food, the bird often takes the shortest route. This algorithm is developed based on this behavior. It employs several particles, each of which is viewed as a point in N-dimensional space. According to its understanding of the appreciable solution and by comparing its best value to the best value of the swarm as of yet, each particle continues to accelerate within the search space. Each particle searches in a certain direction, and through interaction, the bird with the best location thus far seeks to reach that place by modifying their velocity, which requires intelligence. This is perfectly characterized by the notion of social interaction [12, 13].

2.5.4 Advantages of PSO over Previous Two Algorithms

Unlike the BFO algorithm, few parameters require adjustment and it is simple to do so. The only algorithm that does not use the survival of the fittest approach keeps the entire population involved. In contrast to GA, PSO is unaffected by the size of the issue. PSO fixes the GA's flaw, which is premature convergence. This just requires two equations, making it straightforward to implement. Even for complex tasks, fewer than 100 iterations are necessary. Given next are the two main equations of PSO algorithm [14]:

Velocity modification equation:

$$v^{k+1} = wv^k + c\ \text{rand} \times \left(\text{pbest} - s^k\right) + c\ \text{rand} \times \left(\text{gbest} - s^k\right) \tag{2.6}$$

Where, v^k = velocity of agent i at iteration k
  w = weighing function
  c_i = weighing factor
  randi = random number between 0–1
  pbesti = p-best of agent i
  sk = current position of agent i at iteration k
  gbesti = g-best of the group

In Equation 2.6, the first term wv_k is inertia component responsible for movement of particle in the direction it was previously heading. "w" has a vital impact on speed if its value less than it speed up the convergence otherwise encourage exploration.

Second term: c rand × (pbest-s^k) is the cognitive component that acts as the particle's memory.

Third term: c rand × (gbest-s^k) is the social component that is the reason why the particle moves to best region found so far by the swarm.

Once the calculation for velocity of each particle is done, then position can be updated using equation of position modification.

Position modification equation:

$$s^{k+1} = s^k + v^{k+1} \tag{2.7}$$

Where, s^{k+1}, s^k are modified and current search points, respectively
  v^{k+1} = modified velocity

This process is repeated unless and until some stopping criteria are fulfilled.

2.6 PSO-Based Controller Design

Steps (see Figure 2.3):

1. The initial particles are set to some linear position in the choice of Kp and Ki.
2. Their velocities are set to zero.
3. Initial ITAE is set to several values.
4. Evaluate the ITAE for the particles at their matching positions.
5. Initialize pbest for each particle.
6. Find gbest based on minimum ITAE.
7. Start iteration 1.
8. Update the positions.
9. And calculate ITAE at their corresponding position.
10. As a result update pbest and gbest based on ITAE.
11. Update velocity.
12. Iteration=iteration+1.

```
                        ┌─────────────┐
                        │    Start    │
                        └──────┬──────┘
                               │
              ┌────────────────▼─────────────────┐
              │ Select the parameters of PSO;     │
              │ population size,maximum no.of iter│
              │ c₁,c₂ and weights                 │
              └────────────────┬─────────────────┘
```

Start

Select the parameters of PSO; population size,maximum no.of iter c_1,c_2 and weights

Initialization of components(PI gains),each of particles with random position and velocities

Iteration =1 and weight updating

Evaluate the value fitness of each particle in current position

Update p-best and g-best of each particle

Is iter max iter?

Yes

optimal values of PI controller

End

No

Update particles positions and velocities using velocity and position update equations

iter = iter + 1

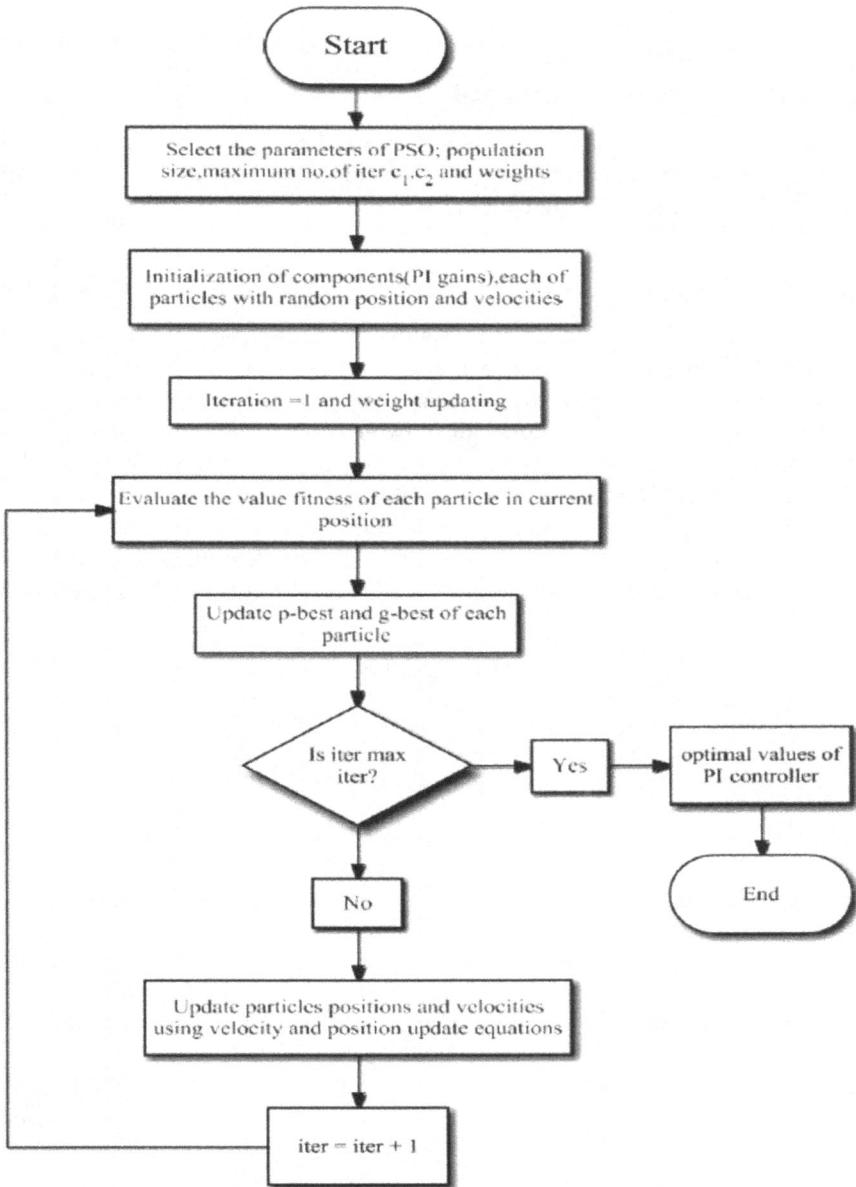

Figure 2.3   Flow Chart of PSO Algorithm.

13. If iteration<=maximum iteration, go to step 8 otherwise continue.
14. The obtained gbest is the optimum set of parameters of PI controller.

2.7 Results and Discussion

At first, PSO algorithm is written taking a general mathematical equation as the objective function to check whether it gives minimum value for the considered equation. The equation considered is given next:

$$x2 + y2$$

Then the result obtained is as shown in Figure 2.4.

And as it is observed, it gives zero as the minimum value for the equation under consideration. This shows that the basic algorithm gave a desirable result.

Then simulation work is done for a two-area interconnected power system according to its block diagram and considering transfer function of each block simulation is done. The parameter values taken in the simulation are tabulated in Table 2.1.

To begin with, the traditional approach is used to determine the kp, ki, and corresponding error values. Similar to that, LQR is also used, and the result is an

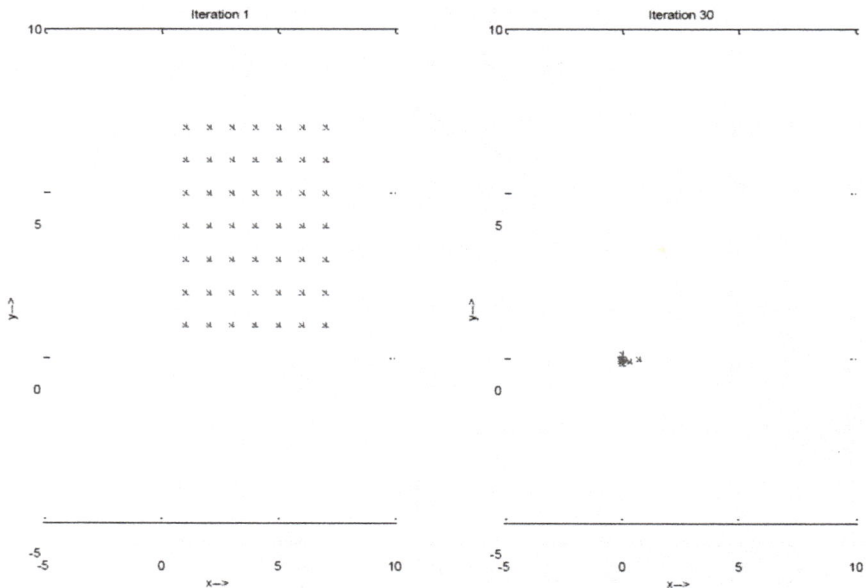

Figure 2.4   Implementation of PSO Taking Objective Function as x2+y2.

Table 2.1   Parameters Used in Simulation

Parameter	Value
B1, B2	0.425 p.u. MW/Hz
R1, R2	2.4 Hz/p.u.
TG1, TG2	0.08 s.
TT1, TT2	0.2 s.
TPS1, TPS2	20 s.
T12	0.0707 p.u.
KPS1, KPS2	120 Hz/p.u.
a12	-1

Table 2.2   Parameter Values Tuned for PSO Algorithm

Parameters	Values
Population Size	11
Numbers of Iterations	50
Inertia Weight (w)	0.8
Cognitive Coefficient (C1)	2
Social Coefficient (C2)	2

Table 2.3   Error Values for Corresponding Methods

Method	Kp	Ki	Error
Conventional	0.60	-0.75	7.6
LQR	Optimal K is given next		2.8
PSO	0.0563	-0.7302	0.7

optimal K value and an error related to it. The PSO method is then used to determine the controller parameter value with the lowest ITAE value. For this, the values in the Table 2.2 are used in tuning. Table 2.3 gives the values of parameters and error for each method applied. Figure 2.5 shows the error obtained by conventional method, LQR and PSO.

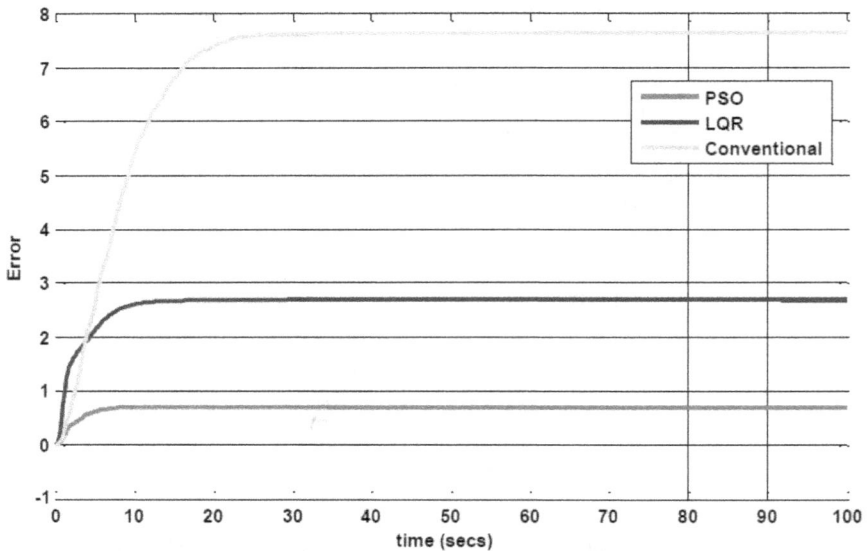

Figure 2.5   Error Obtained by Three Used Methods.

Where, the optimal K is given in the next matrix:

$$K = \begin{bmatrix} 0.5604 & 0.5724 & 0.2057 & 0.2271 & -0.1919 & -0.0644 & 1.0027 & 0.9986 & -0.3376 \\ -0.2450 & -0.1972 & -0.0644 & 0.6029 & 0.5917 & 0.2121 & -1.9131 & 0.3376 & 0.9986 \end{bmatrix}$$

In Figure 2.5, errors of the three methods are compared and shown in the plot.

Simulations in the time domain for step load changes at different locations and with variable parameters are carried out to demonstrate the robustness of the described controller. The replies were improved utilizing a PI controller and PSO with an ITAE objective function.

The following cases are considered:

Case 1: A step load change in area-1 only

A step load 10% rise in area-1 (ΔPD1) is given and the deviation in frequency of area1 Δf1, the deviation in frequency of area-2 (Δf2), and the tie-line power signal of the system. Frequency Deviation is shown in Figure 2.6 for area 1 when applied LQR and PSO.

Case 2: Step load change in area-1 and area-2 simultaneously

In this case, 10% step load rise in demand of first area and 15% step load rise in demand of second area, respectively, are applied. Figure 2.7 shows the Change in Frequency of Area-1 for 0.1 p.u. The response of the system is

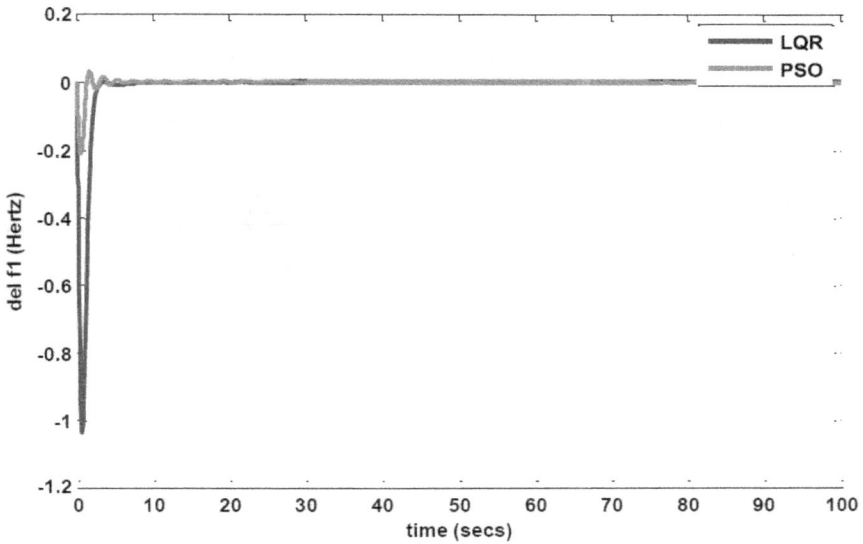

Figure 2.6   Frequency Deviation of Area-1 by LQR and PSO Methods.

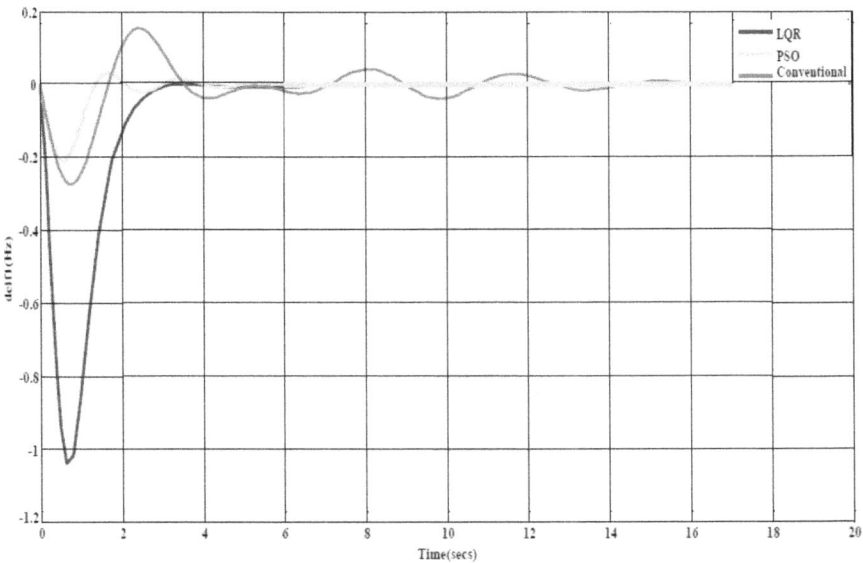

Figure 2.7   Change in Frequency of Area-1 for 0.1 p.u Change in Area-1.

shown in Figure 2.8. Figure 2.9 shows the change in frequency of second area for 0.1 p.u and 0.15 p.u for area-2. Figure 2.10 shows a change in tie line for 0.1 p.u in Area-1 and 0.15 p.u for Area-2.

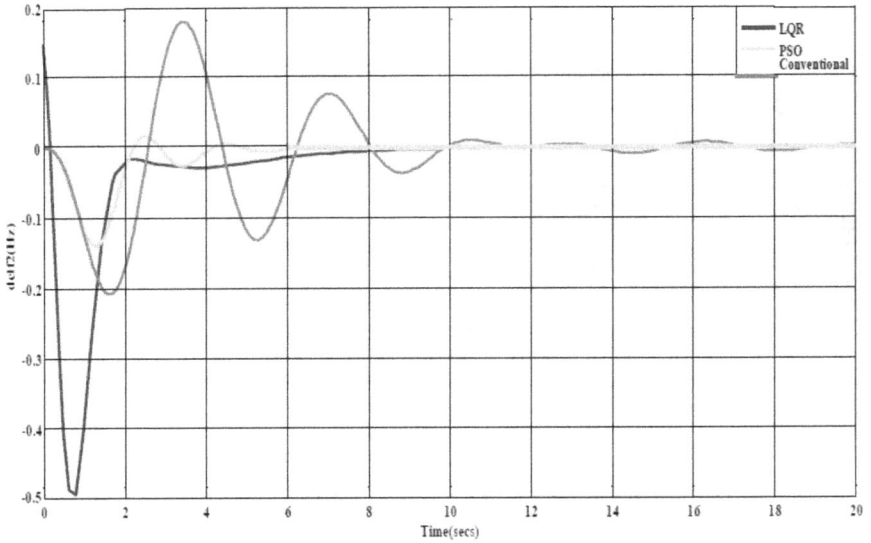

Figure 2.8 Change in Frequency of First Area for 0.1 p.u Change in Area-1 and 0.15 p.u for Area-2.

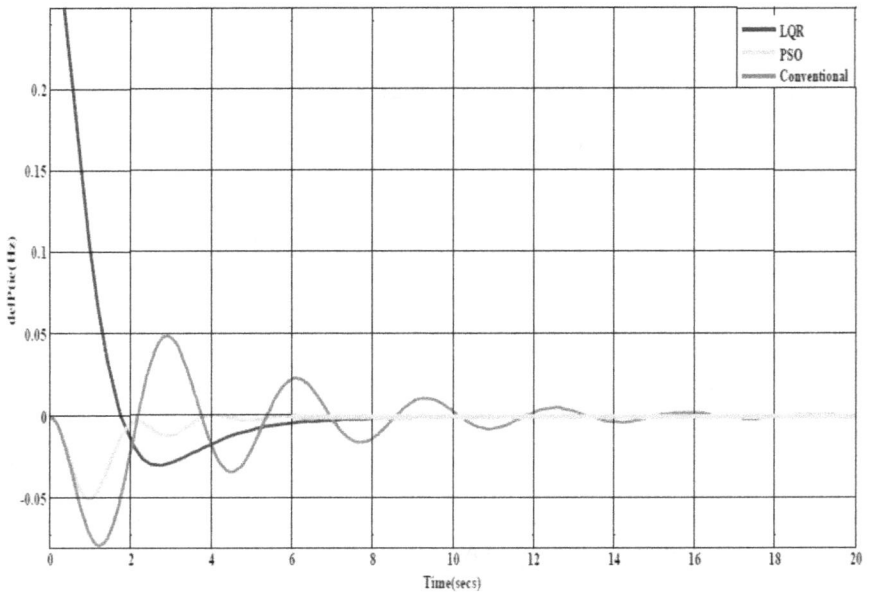

Figure 2.9 Change in Frequency of Second Area for 0.1 p.u Change in Area-1 and 0.15 p.u for Area-2.

Case 3: Robustness of system parameter

To find the robustness of the system TG, TT and T12 are changed by ±30% with frequency deviation of 0.1 p.u in area-1 and 0.15 p.u in area-2 as shown in Figures 2.11–2.13. These next plots show the performance of the controller irrespective of the variation of the time constants of the systems included in these.

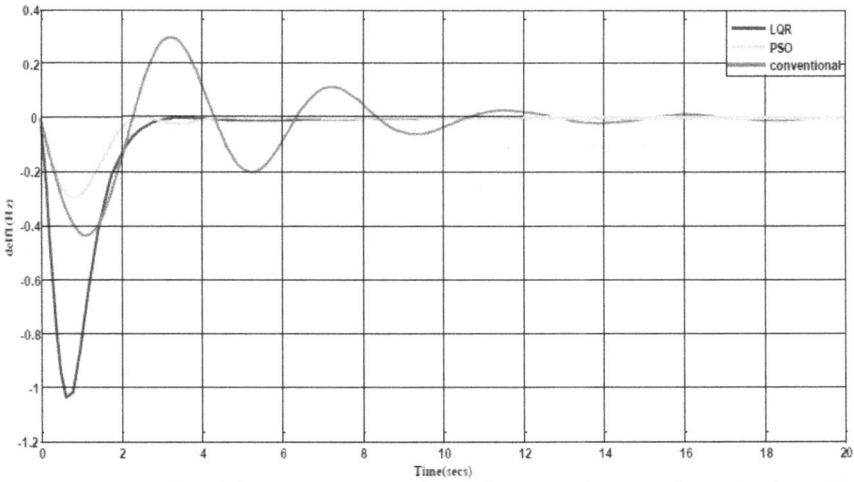

Figure 2.10   Change in Ptie for 0.1 p.u Change in Area-1 and 0.15 p.u Change for Area-2.

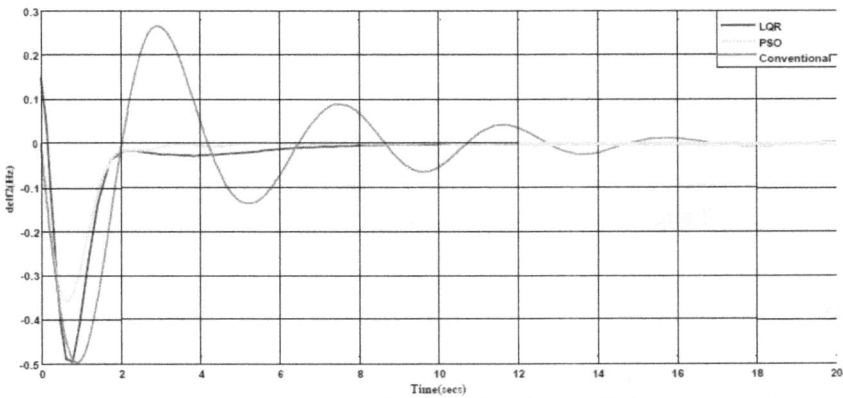

Figure 2.11   Change in Frequency for Change in TG of PI Controller.

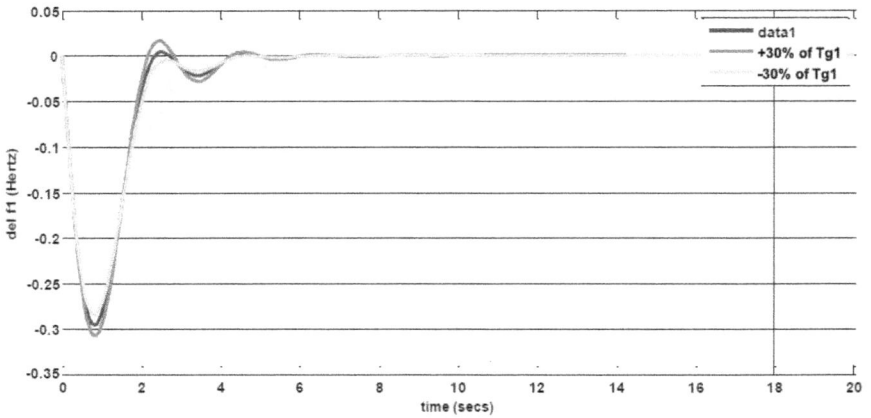

Figure 2.12   Change in Frequency for Change in TT of PI Controller.

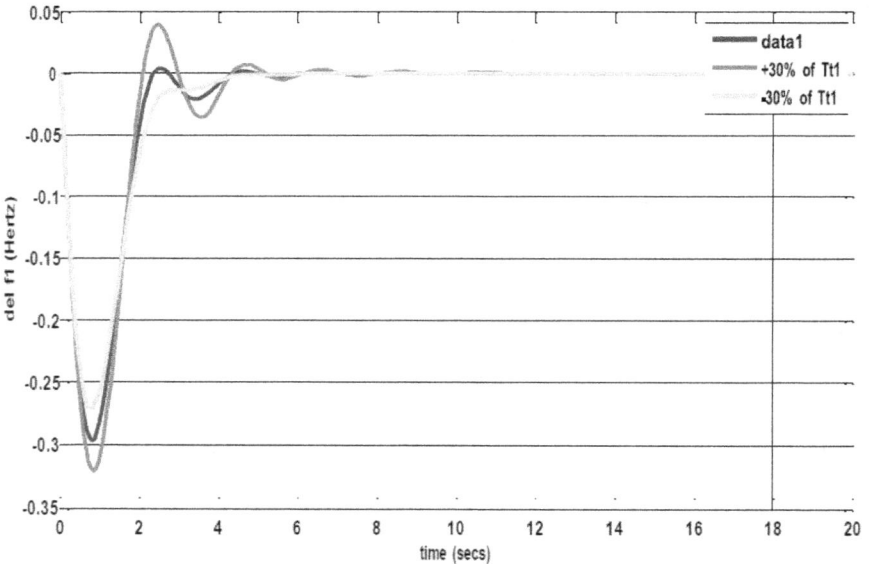

Figure 2.13   Change in Frequency for Change in T12 of PI Controller.

2.8 Conclusion

Designing the best controllers is motivated by the tough challenge of controlling electricity systems to suit consumer needs. They should be able to quickly maintain frequency and voltage while keeping an eye on the power supply. The creation of controllers involves the application of numerous optimization strategies. PSO is employed in this chapter to fine-tune the proportional-plus-integral controller's parameters. The technique is demonstrated using a two-area system. The objective function was the absolute error multiplied by the integral of time. By changing the load requirement of the locations, various frequency deviation charts were produced. Plots and observations were also made of how parameter variation affected the system response. Comparing the error values demonstrates its superiority to alternative controller tuning techniques.

References

1. C. Concordia, L.K. Kirchmayer, "Tie-Line Power & Frequency Control of Electric Power Systems- Part II" AIEE Transactions, Vol. 73, part III-A, 1954, 133–141.
2. O.I. Elgerd, "Energy Systems Theory: An Introduction" McGraw-Hill, New York, 1982. Electric.
3. David Beasley, David R. Bull, Ralph R. Martin, "An Overview of Genetic Algorithms" University Computing, Vol. 15, No. 2, 1993, 58–69.
4. David Beasley, David R. Bull, Ralph R. Martin, "An Overview of Genetic Algorithms: Part 2, Research Topics" University Computing, Vol. 15, No. 4, 1993, 170–181.
5. E.S. Ali, S.M. Abd-Elazim, "BFOA Based Design of PID Controller for Two Area Load Frequency Control with Nonlinearities" Electrical Power and Energy Systems, Vol. 51, 2013, 224–231.
6. R.C. Eberhart, J. Kennedy, "A New Optimizer Using Particle Swarm Theory" Proceedings of the Sixth International Symposium on Micro-Machine and Human Science, Nagoya, Japan, pp. 39–43, 1995.
7. K. Ogata, "Modern Control Engineering" Prentice Hall, Upper Saddle River, NJ, 2008.
8. Ala Eldin Awouda, Rosbi Bin Mamat, "New PID Tuning Rule Using ITAE Criteria" International Journal of Engineering (IJE), Vol. 3, No. 6, 2010.
9. H. Saadat, "Power System Analysis" McGraw-Hill, New York, 1999.
10. I.J. Nagrath, M. Gopal, "Control System Engineering" Fifth Edition, New Age International Publisher, New Delhi, 2007.
11. Allen J. Wood, Bruce F. Wollenberg, "Power Generation Operation and Control" John Wiley & Sons, New York, 1996.
12. Yuhui Shi, Russell Eberhart, "A Modified Particle Swarm Optimizers" IEEE, Indianapolis, IN 46202-5160, 1998.
13. James Blondin, "Particle Swarm Optimization: A Tutorial" September 4, 2009.
14. Deepyaman Maiti, Ayan Acharya, Mithun Chakraborty, "Tuning PID and pIλDδ Controllers Using the Integral Time Absolute Error Criterion" IEEE, New Dehli, 2008.
15. C.L. Wadhwa, "Electrical Power System" Sixth Edition, New Age International Publisher, New Delhi, April 2018.
16. Jeevithavenkatachalam, S. Rajalaxmi, "Automatic Generation Control of Two Area Interconnected Power System Using PSO" IOSR-JEEE, Vol. 6, May 2013.
17. P. Kundur, "Power System Stability and Control" McGraw-Hill, New York, 1994.

Chapter 3

Power Quality Enhancement by PV-UPQC for Non-Linear Load

Eknath Borkar and Nagendra Singh

Contents

DOI: 10.1201/9781003301820-3

3.1 Introduction

The idea of energizing and establishing electric types of gear to guarantee appropriate activity of the supplies according to Institute of Electrical and Electronics Engineers (IEEE) Standard 1159–1995 (IEEE Std 519, 1995) is characterized as power quality (PQ). Universal Electro-specialized Commission (IEC) characterizes PQ as a set of boundaries keeping up the trademark property of flexibly under ordinary working conditions as far as advancement of gracefully like recurrence, extent, waveform, and balance [1]. Extensively two classes of PQ issues are characterized: wonders because of subservient quality of current drawn by the heap brought about by non-direct loads and voltage unsettling influences that cause shortcomings in the force framework [2]. The most huge and basic PQ issues are characterized by transient, voltage dip, voltage swell, harmonics, interruption in voltage, etc. The first published Energy Networks Association Recommendation G5/4 in 2001 intends to ensure that the levels of harmonics in the electricity supply of public network do not disband a problem for other users of that supply. G5/4 is a regulatory recommendation for distribution network used for planning of adding new loads and capacity to an existing utility [3, 4]. While expanding an installation, there is a restriction to limit harmonic emissions for non-linear loads and generation plant onto the electrical supply system. This requirement, if accomplished, will limit voltage distortion in the network connected to non-linear loads, which are the primary factor of high harmonics in the distribution network. G5/4 specifies the limits of harmonics, that is, THD 4% for 6.6 KV to 20 KV upto 50th order and 8% for 33 KV to 35 KV in the network to facilitate the connection of non-linear equipment, having manifesto for limiting the overall voltage distortion, which in turn are set to achieve compatibility to the network against PQ issues [2]. Innovative solutions are rapidly evolving, which employ power electronic devices (PEDs) for quick response to suppress or counteract the disturbances. There are two general approaches to mitigate the PQ problems [3]. One is load conditioning to ensure that the connected equipment is less sensitive to disturbances, which can allow ride through the disturbances. The other is the installation of conditioning devices to ensure suppression or counteract the PQ issues. Commercially available PED tends to shield against a group of PQ problems. These devices vary in size and can be installed at all voltage levels of a power system [4–6] (Figure 3.1).

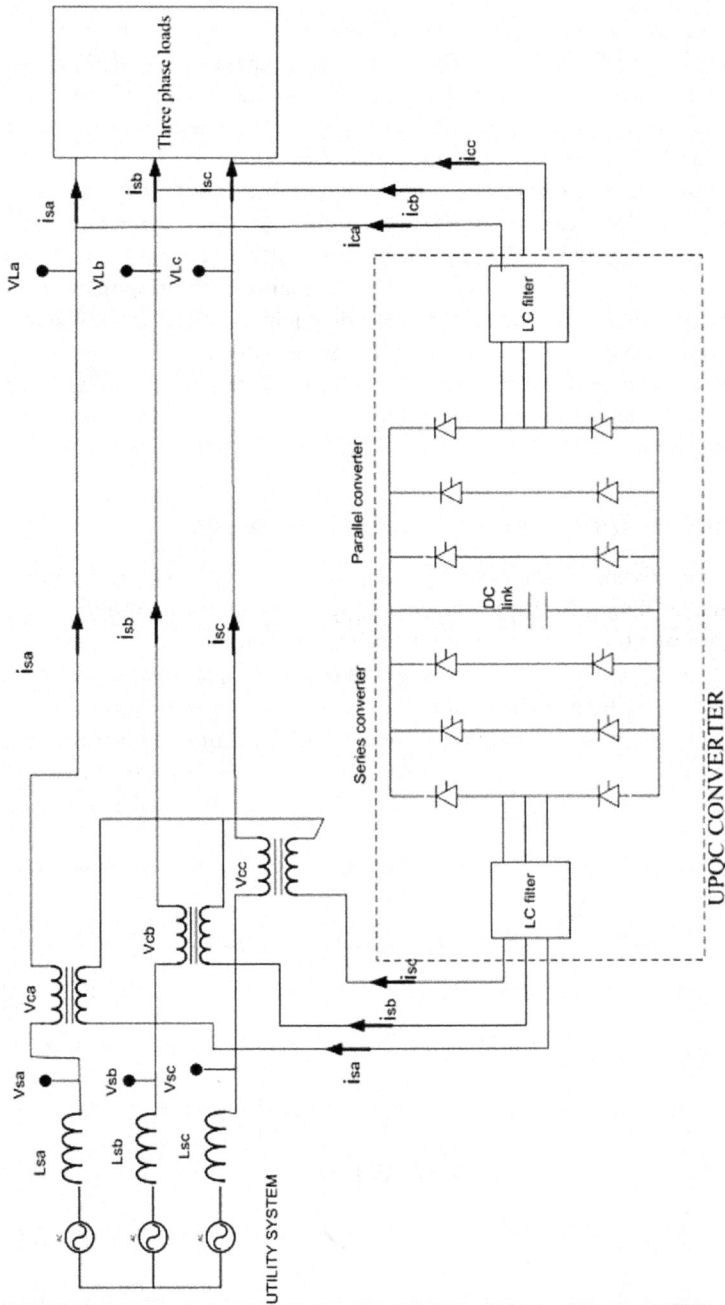

Figure 3.1 Schematic of UPQC.

3.2 Unified Power Quality Conditioner (UPQC)

UPQC is composed of two voltage source inverters. It can simultaneously perform the tasks of APF and DVR [7, 8]. The basic circuit diagram of UPQC is shown in Figure 3.2. UPQC protects the loads against voltage sag, swell, voltage unbalance, harmonics, and low power factor. UPQC is a combination of a shunt-series connected APF and DVR respectively inter-linked by a common DC link capacitor [9, 10]. The processing of power at supply, load and in a way to compensate reactive and harmonic by means of PEDs is becoming more prevalent due to the vast advantages offered by them [11–13]. The shunt APF is typically associated over the heaps to relieve all current related issues, for example, the responsive force pay, power factor improvement, current symphonies pay, and burden unbalance pay, while the arrangement dynamic force channel is associated in an arrangement with a line through arrangement transformer. It goes about as controlled voltage gracefully and can repay all voltage related issues, for example, voltage sounds, voltage hang, voltage swell, and so on.

3.2.1 Control Techniques of UPQC Topologies

The control unit forms an important measure of the UPQC system. Fast detection of disturbance in signals having high accuracy, processing of the reference, and high dynamic response of the controller are the prime necessities for desired UPQC [14]. The main deliberations for the control system of a UPQC include series control: sag/swell recognition, voltage reference age, voltage infusion systems, and techniques for creating of gating signals; and shunt inverter control: current reference age, strategies for producing of gating signs, and capacitor voltage control [15, 16]. See Figure 3.3.

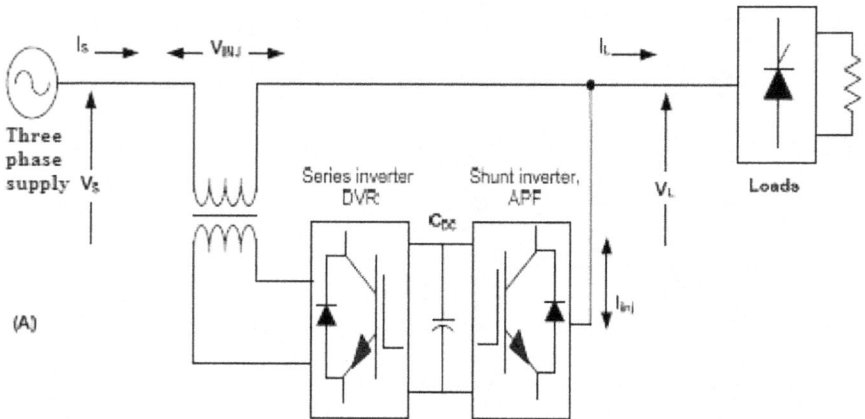

Figure 3.2 Circuit Topology of UPQC.

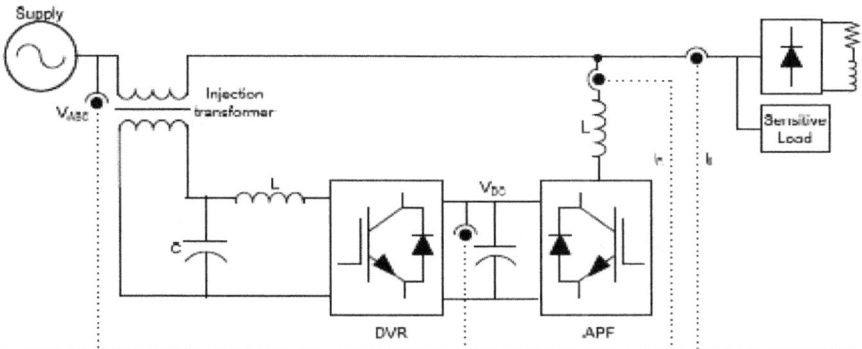

Figure 3.3   Control Unit of UPQC.

3.2.2 PV Fed UPQC (PV-UPQC)

In the distribution system, the PQ issues are there on account of non-linear loading and grid connected-system, so recently versatile UPQC controller is preferred [17–21]. An UPQC is to energize it via PV-DC power to increase the green element into the network connected. Integrating PV-array to UPQC has put together dual technical benefits of clean energy generation along with universal active filtering [22, 23]. This unique combination of green source with PED that is PV-UPQC has been reported in [24–26]. Compared to conventional CPDs mentioned in this chapter before, PV-UPQC has several benefits such as improving PQ of the grid, protecting critical loads from grid side unsettling influences separated from expanding the issue ride through capacity of converter during homeless people likewise forestalling non-linearity of burden yet to interfere with the source [27, 28]. With the expanded accentuation on conveyed age and small-scale frameworks, there is a reestablished enthusiasm for UPQC frameworks [29, 30]. The PV-UPQC can perform two functions. First, the PV-array can energize the electrical distribution system. And second, it can provide power only to the load, similar to an uninterruptible power supply [31, 32].

3.3 Schematic of PV-UPQC

This work eliminates the PQ issues generated due to non-linear loading and grid connected PV-system. A versatile UPQC controller is preferred in the distribution system. Photovoltaic (PV)-tied Unified Power Quality Conditioner (UPQC) are connected back-to-back through DC-link capacitor and have been reported for simultaneous mitigation of both current related and voltage-based power quality issues [33]. The UPQC has its own supply system that is energized through the PV system. The PV-UPQC can mitigate PQ related issues and can integrate PV with the utility system. A strategy with duality in compensation is adopted to operate the

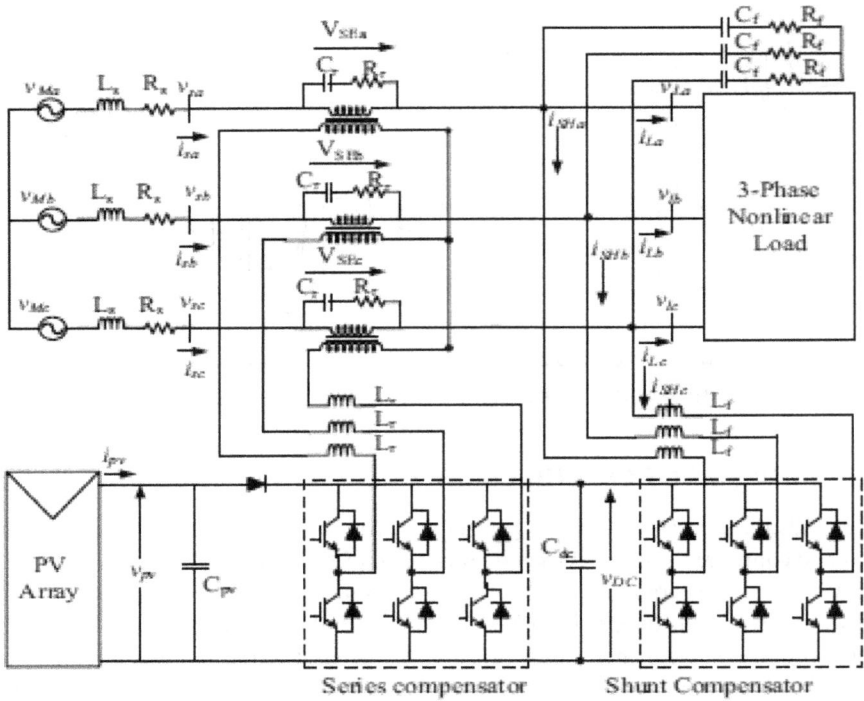

Figure 3.4   Schematic PV-UPQC System.

PV-UPQC system, where the shunted converter is controlled as a sinusoidal voltage source and the series converter is controlled to operate as a sinusoidal current source [34, 35]. The proposed system is shown in Figure 3.4.

3.4 System Configuration and Design

A MATLAB® Simulink® model of the solar panel has been developed whose DC output is regulated using DC-DC boost converter. A three-phase voltage source is considered an AC bus, replica of grid with short circuit capacity of 100MVA.

The basic functionality of UPQC is to maintain grid profile at all adverse operating conditions like non-linear loading, unbalanced loading or the condition of voltage fluctuations. UPQC has its own source to energize the converters connected. Generally, this power source is DC-batteries. In the work done, these batteries are energized using solar power, hence making it eco-friendly.

■ A UPQC controller is designed, which is composed of two converters: One is in series, which operates as a sinusoidal current source; another is in parallel operating as a sinusoidal voltage source [36, 37].

- Impedance of arrangement converter must be sufficiently high to confine the symphonious flows created by the non-linear burdens.
- Impedance of parallel converter must be sufficiently low to absorb the load harmonic currents.
- The converter is designed using Enhanced PLL and PI controller, and an analogue low-pass filter is designed to mitigate the system harmonics.
- A conventional three-leg full bridge universal inverter is used. To generate the gate pulses for inverter, 2-level PWM is used whose output is controlled by PI. The PI is referenced using grid current. To reduce complexity, an abc-dq transform is used in a three-phase system. The simulation model and output waveforms of all the three modes of operation are presented in the subsequent sections.

For a three-phase system to PQ issues caused by non-linear loading, unbalanced loading, or the condition of voltage fluctuations, a PV-UPQC system is designed using dual compensating structure of series-shunt converters. The three main components of the designed system are series converter, shunt converter, and PV-Array.

3.5 Controller Design

A UPQC controller is designed, which is composed of two converters: One is in series that operates as a sinusoidal current source; another is in parallel operating as sinusoidal voltage source. Impedance of series converter must be high enough to isolate the harmonic currents generated by the non-linear loads. Impedance of parallel converter must be confined to low in a way of absorbing the load harmonic currents. The converter is designed using Enhanced PLL and PI controller and an analogue low-pass filter is designed to mitigate the system harmonics. A conventional three-leg full bridge universal inverter is used. To generate the gate pulses for inverter, 2-level PWM is used whose output is controlled by PI. The PI is referenced using grid current. For a three-phase system to reduce complexity, an abc-dq transform is used as shown in Figure 3.5. The Kp and Ki gain for PI is 0.01 and 500.

3.6 Filter Design

A low-pass filter (LPF) is a channel that passes signals with a recurrence lower than a selected cut-off recurrence and weakens signals with frequencies higher than the cut-off recurrence. The specific recurrence reaction of the channel relies upon the channel structure. In a utility system, filters have tremendous utilization since the harmonics present in the voltage and current can be eliminated with the help of filters. The basic configuration of LPF is presented in Figure 3.6.

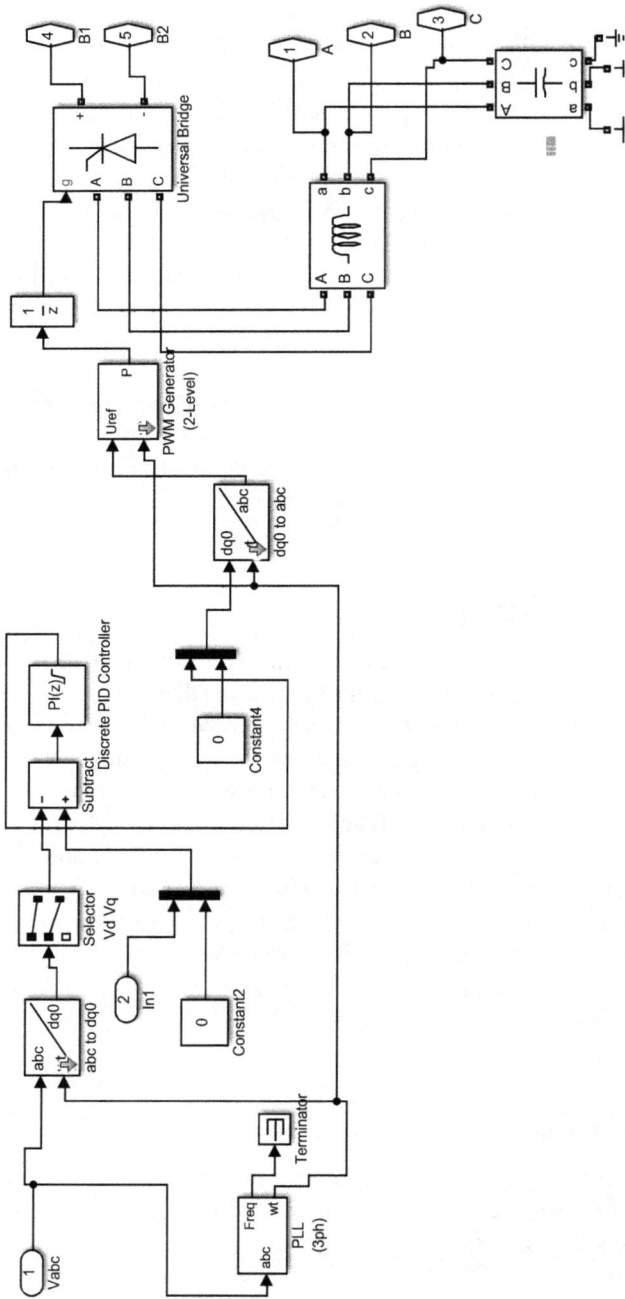

Figure 3.5   PLL-PI Controller.

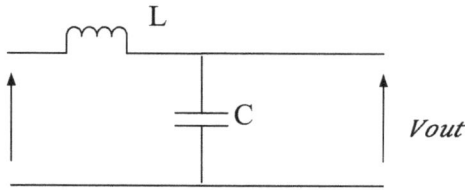

Figure 3.6   Schematic of Analogue LPF.

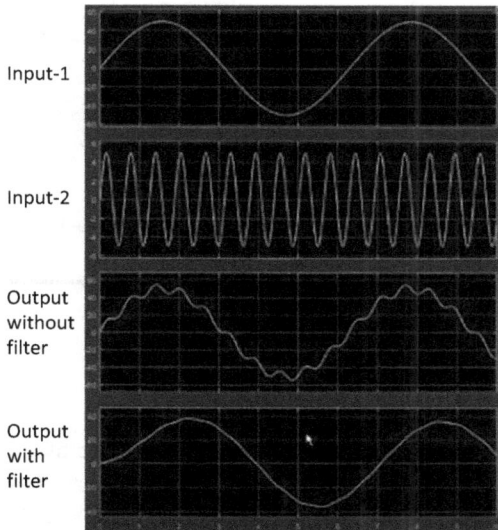

Figure 3.7   Waveforms for Understanding Functioning of LPF.

$$L = \frac{0.03V_{in}}{2\pi f I_{Lmax}} \qquad (3.1)$$

$$C = \frac{1}{(2\pi f)^2 L} \qquad (3.2)$$

To understand the functioning of LPF, consider input-1 with fine sine wave. Another input is fed into the system with different frequency. When both inputs are merged, the output obtained is distorted. But when an LPF is placed between a fine sine wave of voltage, it is obtained again, as shown in Figure 3.7.

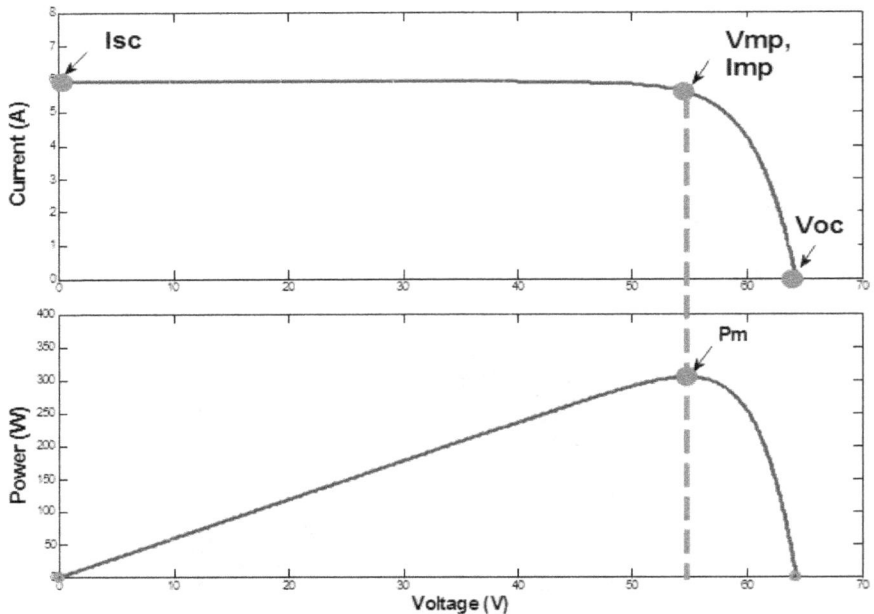

Figure 3.8 VI-PV Characteristics of PV Module.

3.7 PV MODEL

The modeling of PV system is done using the PV module block of MATLAB. The PV module is a preset block browsed from system data base having PV-VI characteristics as shown in Figure 3.8. The open circuit voltage Voc (V) is obtained when array terminals are left open and whose default value is 36.3 V. The short-circuit current Isc (A) is obtained when array terminals are short circuited. Voltage is at maximum power point Vmp (V) with default value of 29 V. Current at maximum power point Imp (A) with default value is 7.35 A. The maximum power is the product of Vmp and Imp with 213.15 W.

3.8 Simulation Studies

3.8.1 Simulation Model

The complete simulation model of the system designed is presented in Figure 3.9. The parameter selection is presented in Table 3.1. A 150 KW PV system with 600 V PV DC output is considered to feed the DC-link capacitor of PV-UPQC system. A three-phase voltage source is considered an AC bus, replica of grid with short circuit capacity of 100MVA.

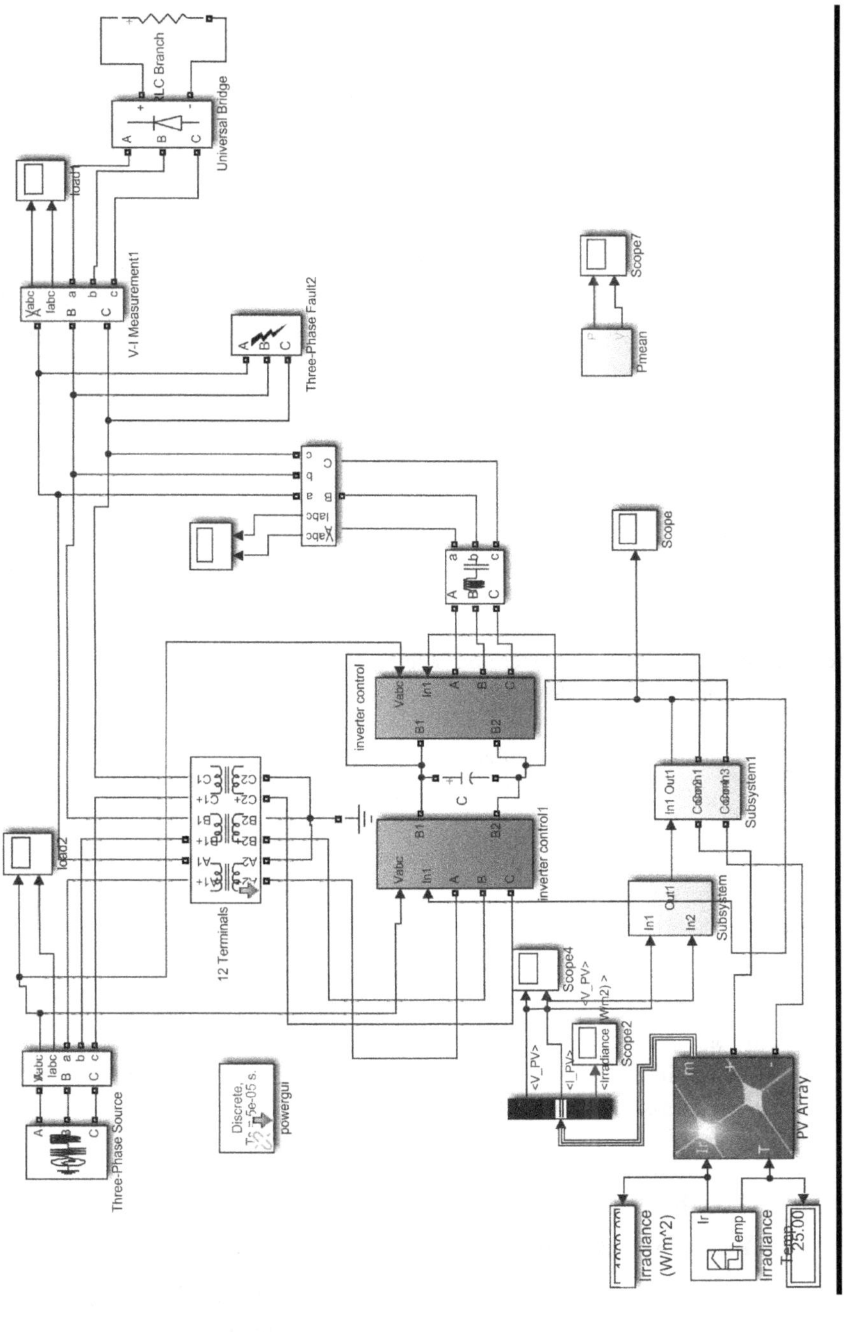

Figure 3.9 MATLAB® Model of PV-UPQC.

Table 3.1 Parameter Selection for the Proposed Work

Parameter	Symbol	Value
Nominal Utility Voltages (rms)	V	415 V
Nominal Frequency	f	50Hz
Inverter Inductance	L	45mH
Filter Capacitance	C	60 μ F
Filter Inductance	L_f	10.45 mH
Linear Load 1	Three-phase resistive load	R=100 ohms
Non-linear Load 2	Three-phase load through rectifier	R=20 ohms
DC-link Voltage	Vdc	900 V

To design a UPQC, two back-to-back DC/AC converter is connected through a DC-link capacitor with 1micro farad capacitance. One side of the converter is connected to the synchronized AC output of the PV system and other side to the grid. The system is synchronized with the grid using PI controller and Phase Lock Loop. The system is analyzed for linear loading of 50 KW and non-linear loading of 40 ohms connected through a three-phase rectifier. The performance analysis of the designed PV-UPQC system under the following three operating modes has been carried out:

1. Performance of PV-UPQC under Varying Irradiation.
2. Performance of PV-UPQC at Load Unbalancing Condition.
3. Performance of PV-UPQC at PCC Voltage Fluctuations.

3.9 Simulation Model of DC Boost Converter

The PV voltage output of solar system is regulated to obtain a fixed DC voltage with the help of a boost converter. A boost converter is a DC converter with input obtained from solar and an inductor and capacitor in series—parallel respectively is connected as shown in Figure 3.10. The output voltage generated is controlled via IGBT, which is triggered through PWM generator.

3.10 Performance of PV-UPQC under Varying Irradiation

Under this condition, non-linear loading of 40 ohms connected through three-phase rectifier. The power output and voltage of PV-module under this condition is given in Figure 3.11 and Figure 3.12 presents the DC-bus voltage. First, to justify the operation of PV-UPQC system the system is analyzed without connecting

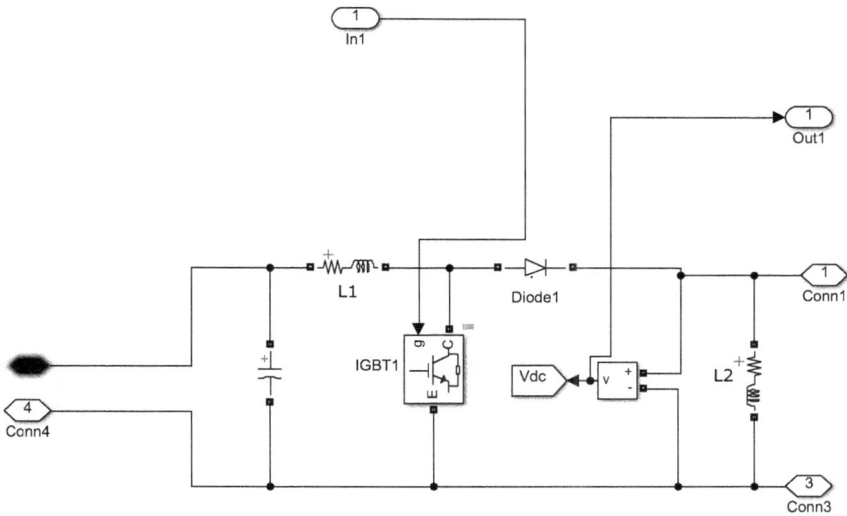

Figure 3.10   MATLAB® Model of Boost Converter.

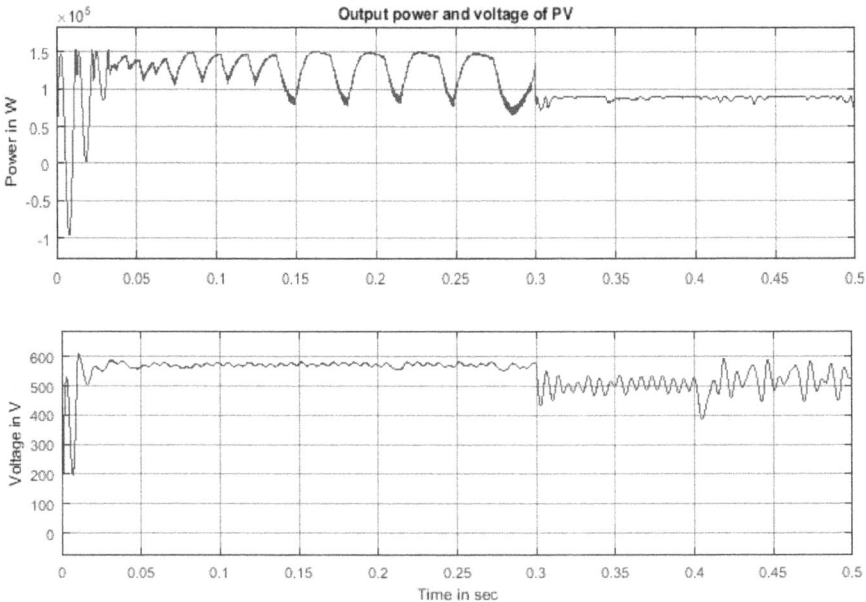

Figure 3.11   Output Power and Voltage of PV at Variable Irradiance.

the PV-UPQC controller. The output voltage and current waveforms source side without PV-UPQC are given in Figure 3.13, and Figure 3.14 presents the output voltage and current waveforms source side with PV-UPQC. The output voltage and current waveforms load side is given in Figure 3.15 without PV-UPQC, and

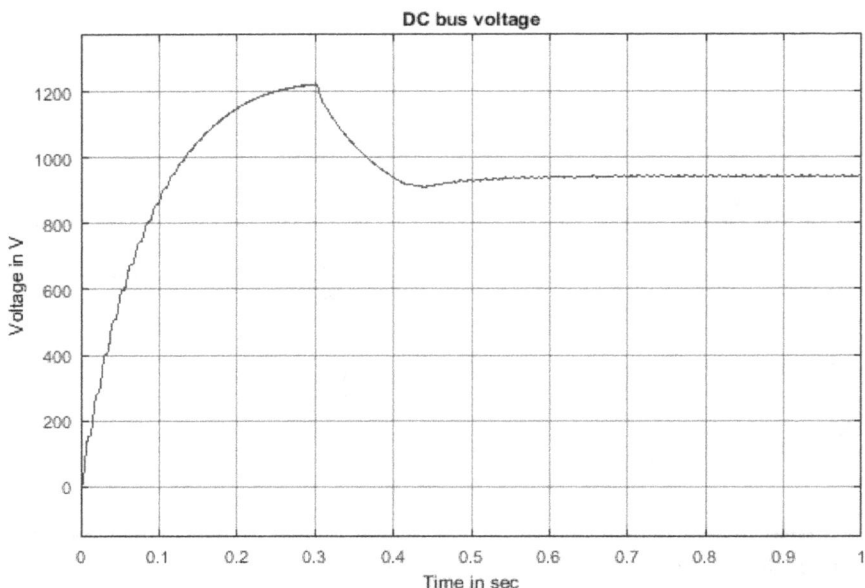

Figure 3.12   Output Voltage of DC-bus at Variable Irradiance.

Figure 3.16 presents the output voltage and current waveforms load side with PV-UPQC.

From the figures of output voltage and current waveforms for both the conditions when PV-UPQC is not connected and when connected, it can be observed that the non-linearity of load is also the propagated source side that can be mitigated using a proposed controller.

From the previous figures, it can be seen that under the condition of non-linear loading, the load voltage and current are distorted and when PV-UPQC is connected at load side, the harmonics in waveforms is reduced to a large extent, hence obtaining a more sinusoidal output.

3.11 Performance of PV-UPQC under Unbalance Loading

Under this condition, a linear load of 40 KW is connected and after time 0.3 sec load of phase b' is disconnected. The system has the same profile both at load side and source side. But while connected PV-UPQC though load voltage and current get unbalanced as shown in Figure 3.17, and source profile is retained as shown in Figure 3.18.

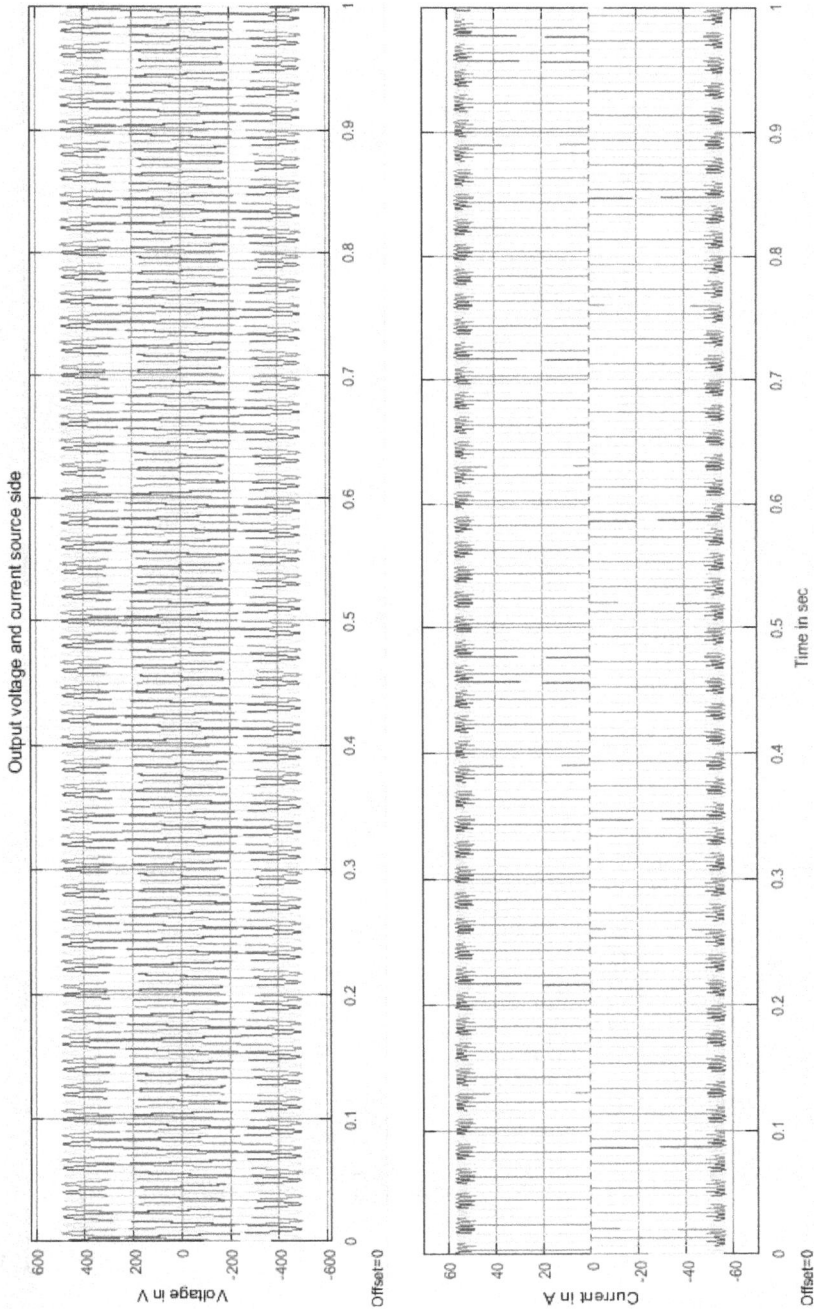

Figure 3.13 Output Waveform Grid Side for Variable Irradiance with Non-Linear Loading without PV-UPQC.

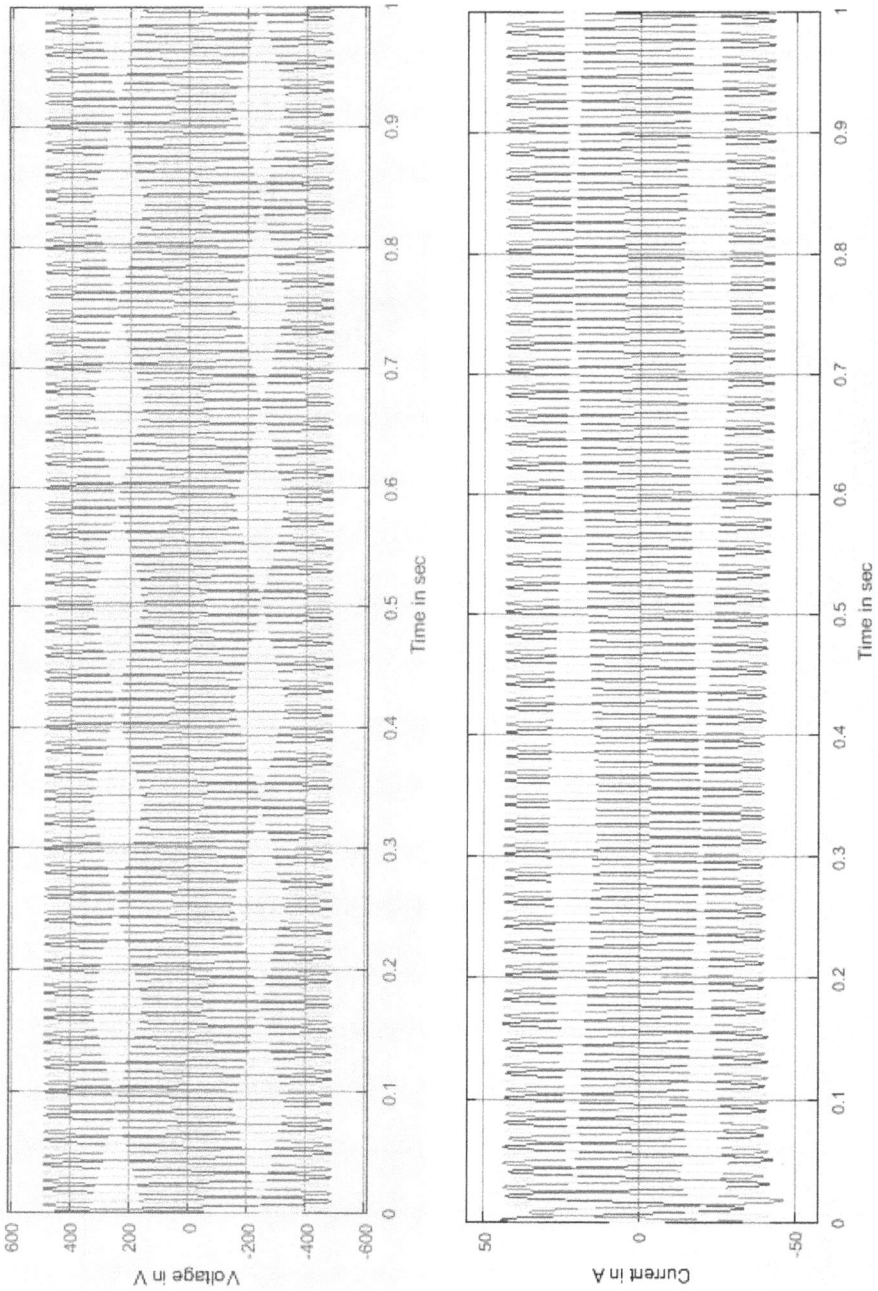

Figure 3.14 Output Waveform Grid Side for Variable Irradiance with Non-Linear Loading with PV-UPQC.

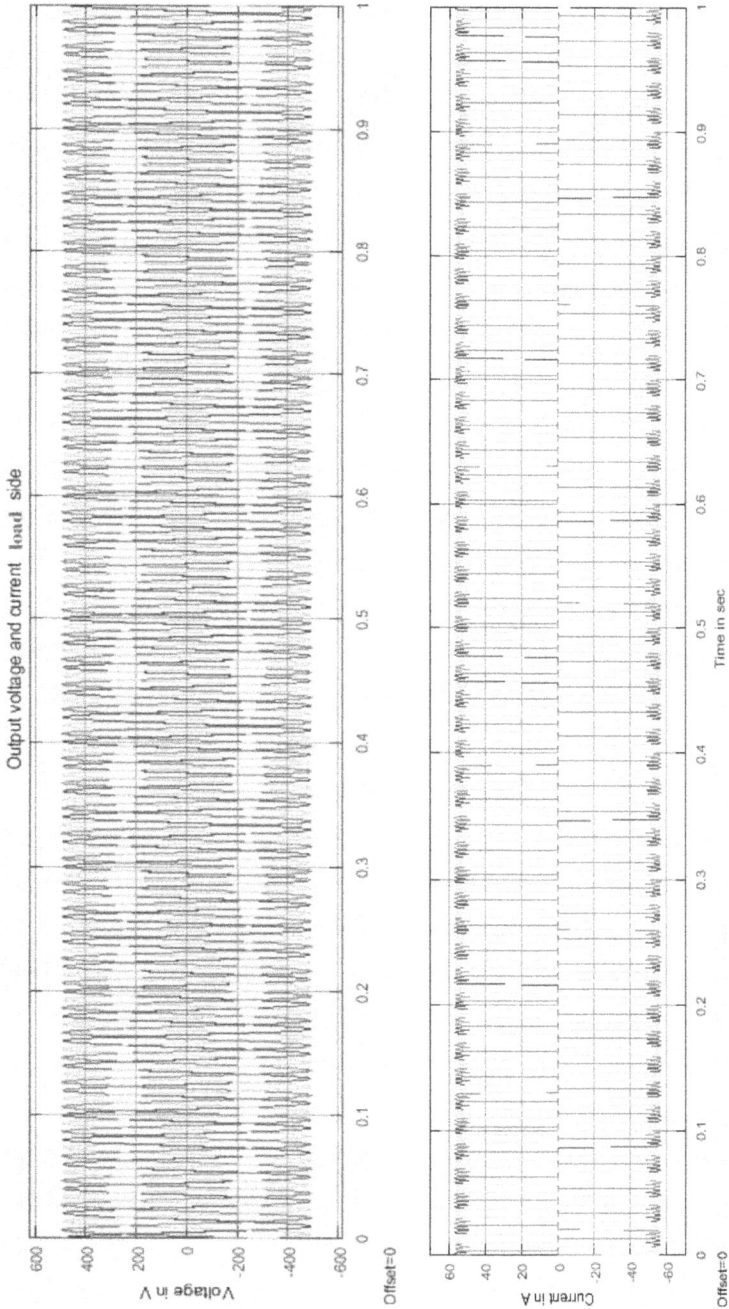

Figure 3.15 Output Waveform Load Side for Variable Irradiance with Non-Linear Loading without PV-UPQC.

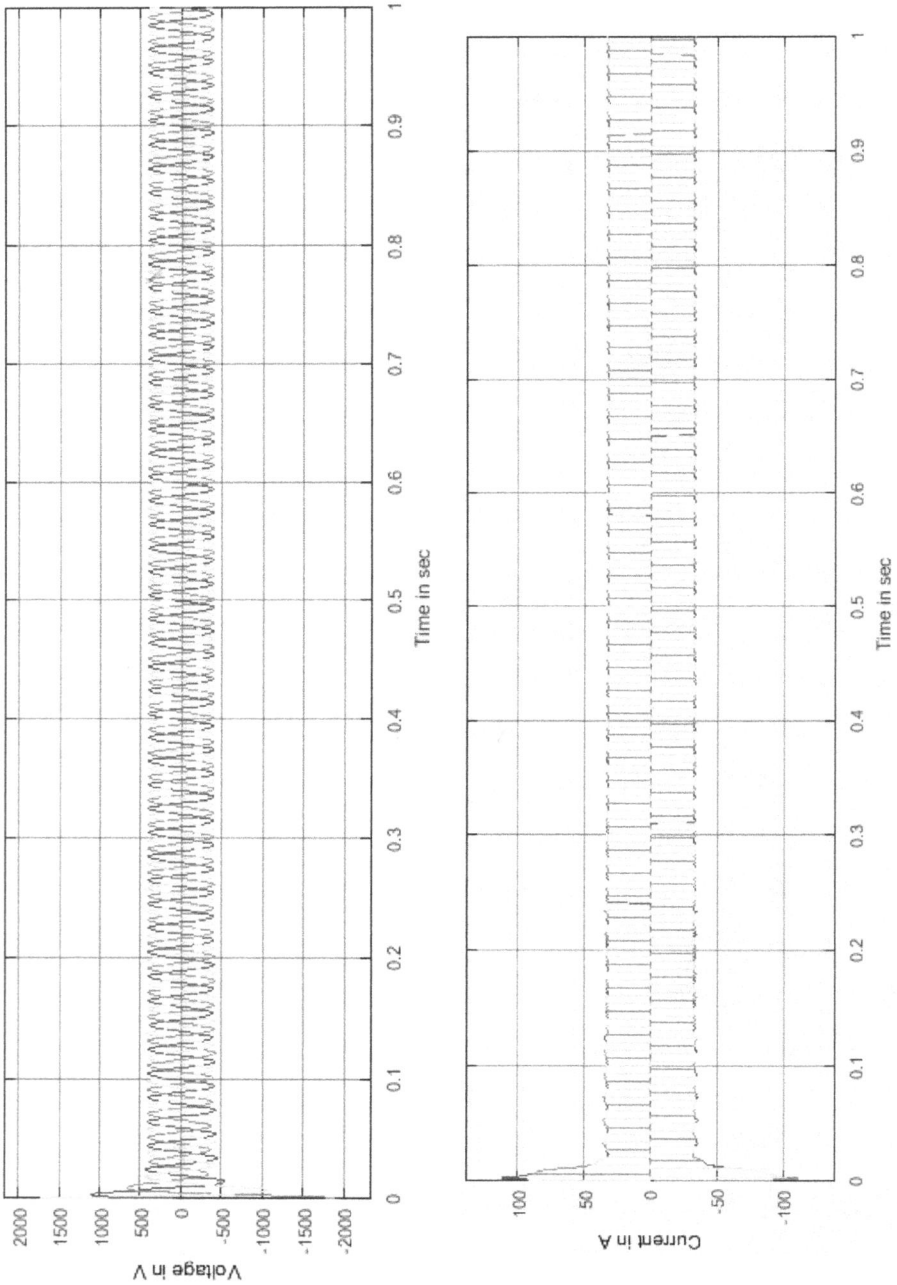

Figure 3.16 Output Waveform of Load Side for Variable Irradiance with Non-Linear Loading with PV-UPQC.

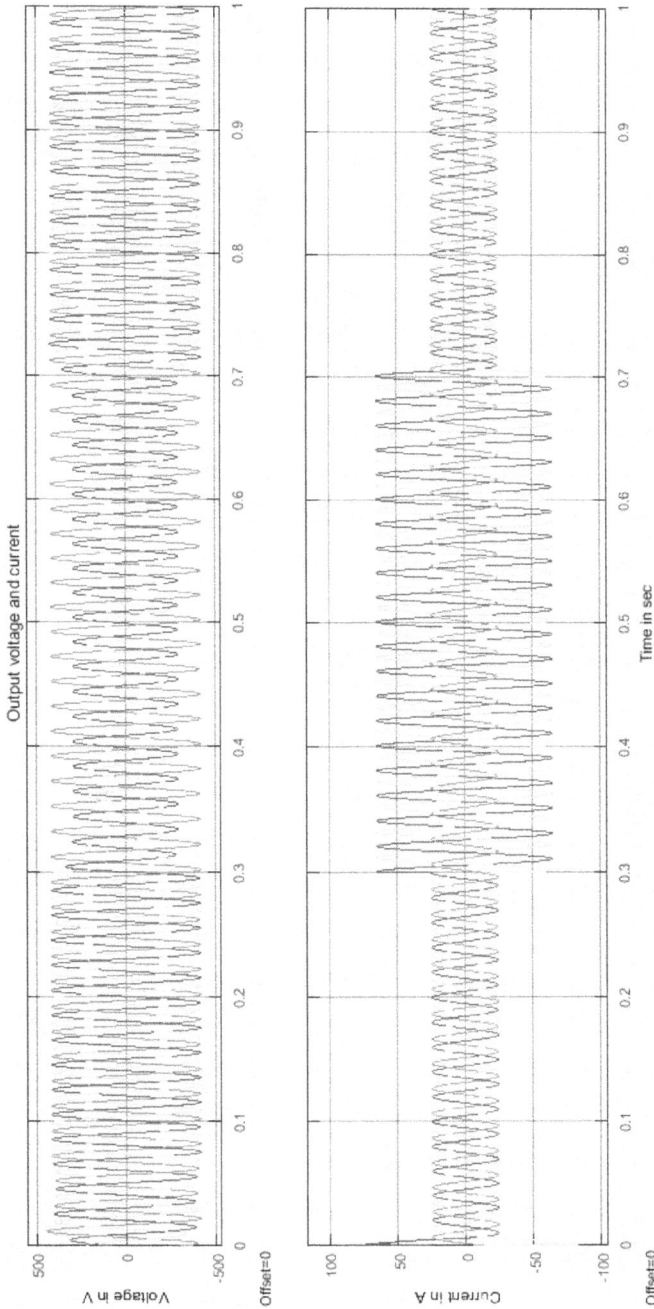

Figure 3.17   Output Voltage at the Condition of Unbalanced Loading at Load Side with PV-UPQC Controller.

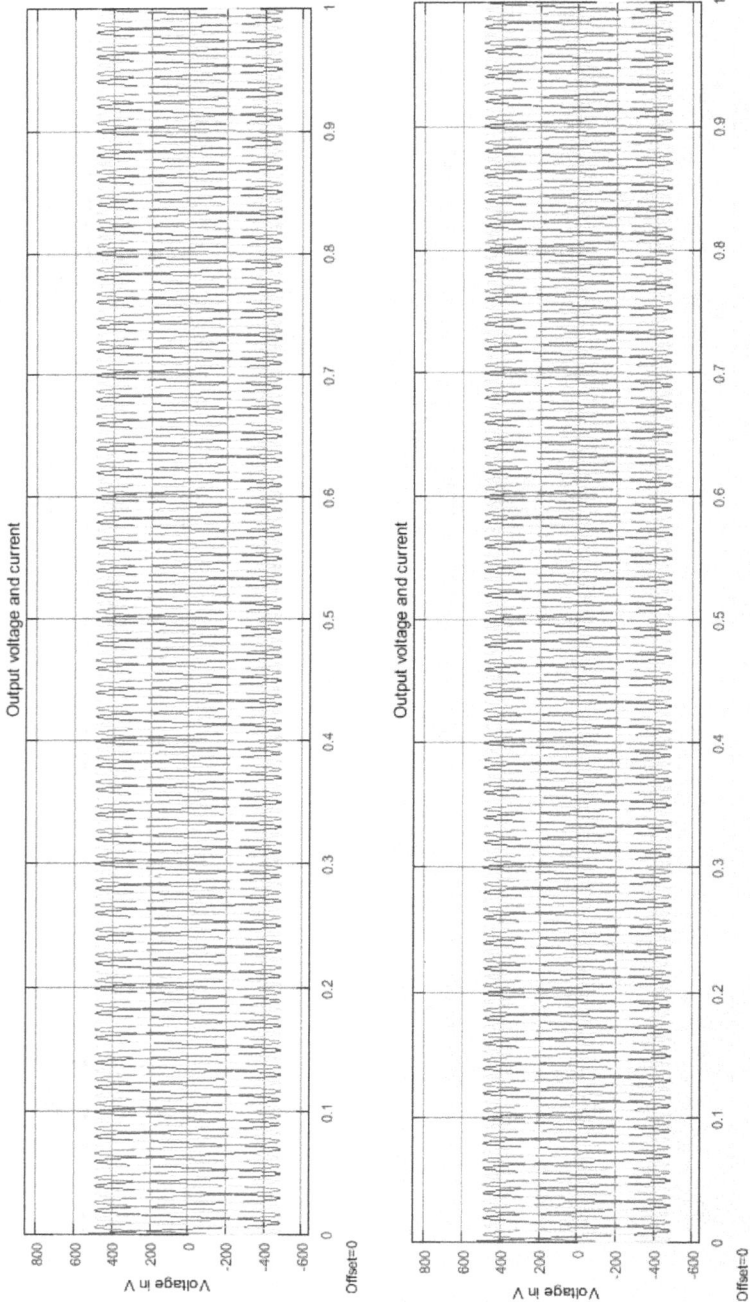

Figure 3.18 Output Voltage at the Condition of Unbalanced Loading at Source Side with PV-UPQC Controller.

3.12 Performance of PV-UPQC at PCC Voltage Fluctuations

In this mode of operation, balanced linear loading is applied to the system to study the behavior of the proposed topology and a condition of voltage fluctuation is created at source side. The output waveforms at grid side and load side are presented in Figure 3.19 and Figure 3.20. When a voltage sag as shown in Figure 3.20 is created, both voltage as well as current at source bus undergoes dip, while load voltages are maintained at constant profile, hence protecting the voltage sensitive equipment connecting to the system.

3.13 Experimental Results

The simulation results for the proposed PV-UPQC system under variable irradiance and unbalanced loading, which may be the result of fault in the distribution network, have been presented in the work done, also comparative analysis THD of with and without PV-UPQC for Load and Source Current is also shown in Table 3.2. Non-linear loading has very high harmonic content. If these harmonics in the load current and load voltages are not mitigated at the load side, they may pollute the source profile. They also can damage the equipment connected to the same source. The THD analysis of voltage waveform load side for voltage sag in Figure 3.21 and the work done is presented in Figure 3.22. The load current harmonics when non-linear load is connected are 21.8% and source current harmonics are 0.88%. From the THD spectrum, it can be seen that the proposed PV-UPQC system successfully mitigated the power quality issues by eliminating source current harmonics and regulating load voltages.

3.14 Conclusion

The UPQC has its own source specifically DC to feed the power electronic devices connected and to maintain the power quality of the system where it is connected. In this chapter, UPQC is fed by PV source, and it has been analyzed for variable irradiation and grid unbalance conditions. The simulation model and performance of the system have been studied in MATLAB-SIMULINK tool kit. It is seen that PV-UPQC mitigates the music brought about by non-linear burden and voltage hang. The framework effectively keeps up the THD of matrix current under constraints of IEEE-519 norm. The stable system for voltage and current profile has been obtained for variations in irradiations, and also mitigation of voltage sags/swell and load unbalance. Table 3.3 shows a comparative analysis of THD with & without PV-UPQC for Load and Source Voltage. The PV-UPQC is a good solution for obtaining green distribution system with power quality improvement. Static and

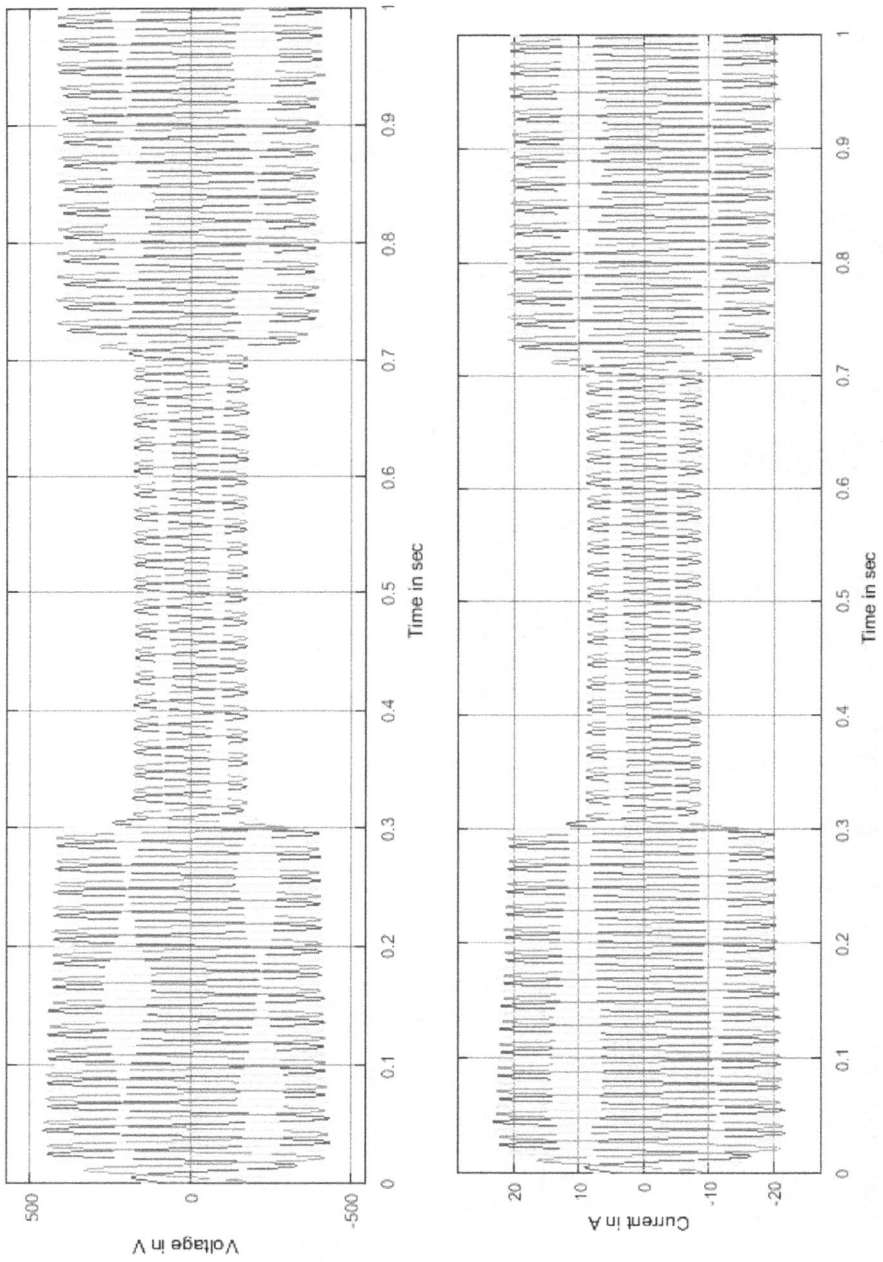

Figure 3.19 **Output Waveform for the Condition of Voltage Sag Source Side.**

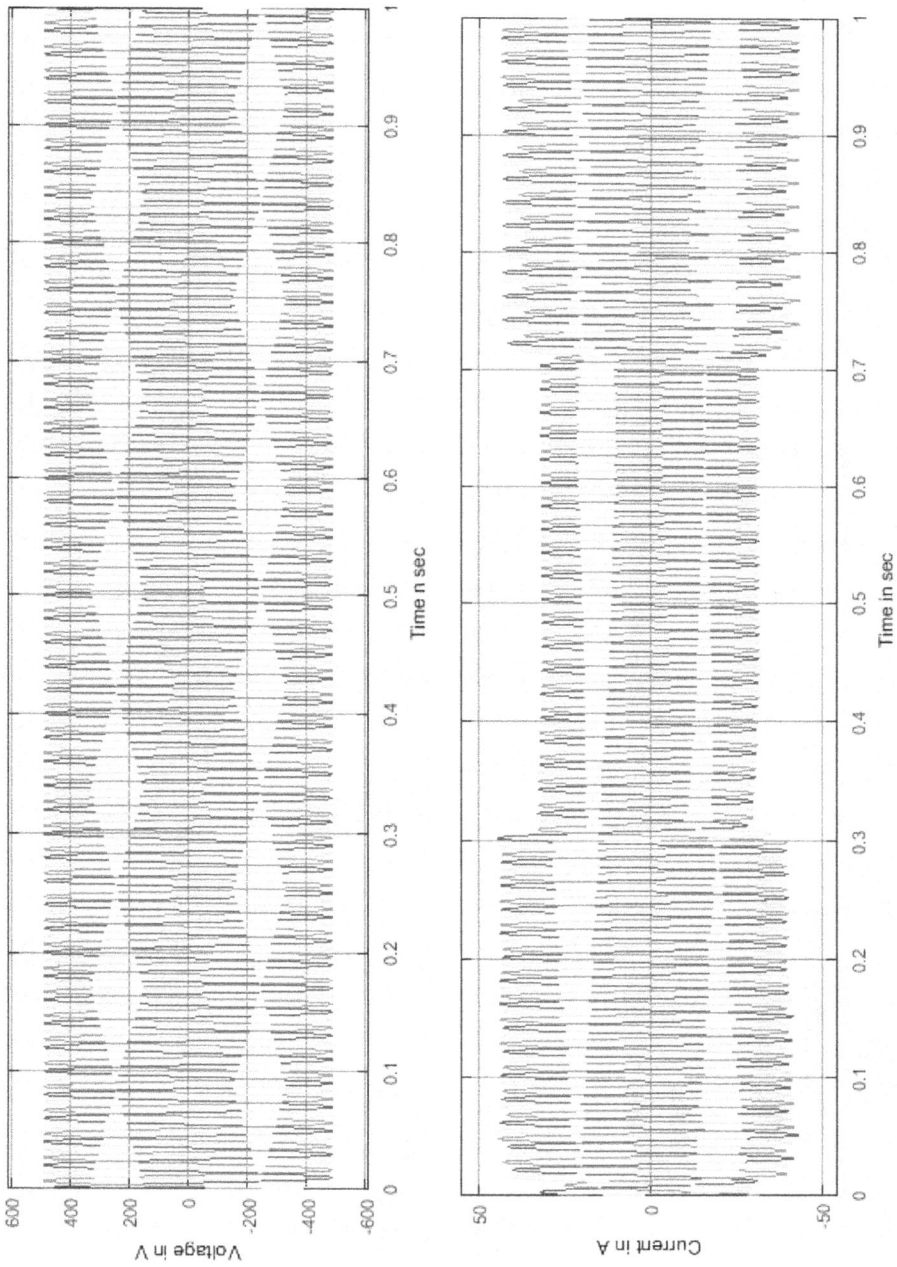

Figure 3.20   Output waveform for the Condition of Voltage Sag Load Side.

Table 3.2   Comparative Analysis THD of with and without PV-UPQC for Load and Source Current

	THD% Load Current					
Conditions	Without UPQC			With UPQC		
Phase Current	I_R	I_Y	I_B	I_R	I_Y	I_B
Case 1	0.07	0.16	0.16	0.07	0.16	0.16
Case 2	28.27	28.27	28.27	21.8	21.8	21.8
Case 3	28.27	28.27	28.27	22.7	22.7	22.7
	THD% Source Current					
	Without UPQC			With UPQC		
Case 1	0.07	0.16	0.16	0.08	0.16	0.16
Case 2	28.27	28.27	28.27	0.88	0.88	0.88
Case 3	28.27	28.27	28.27	1.99	1.99	1.99

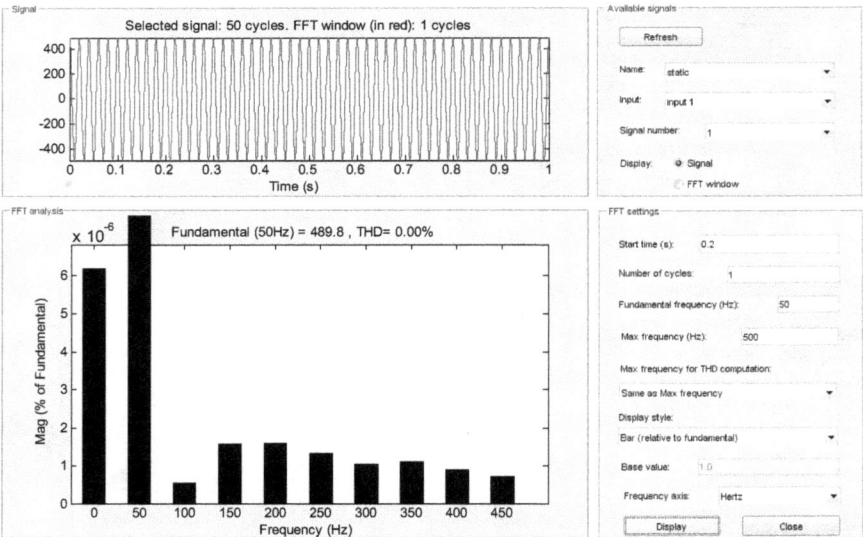

Figure 3.21   THD Analysis of Voltage Waveform Load Side for Voltage Sag.

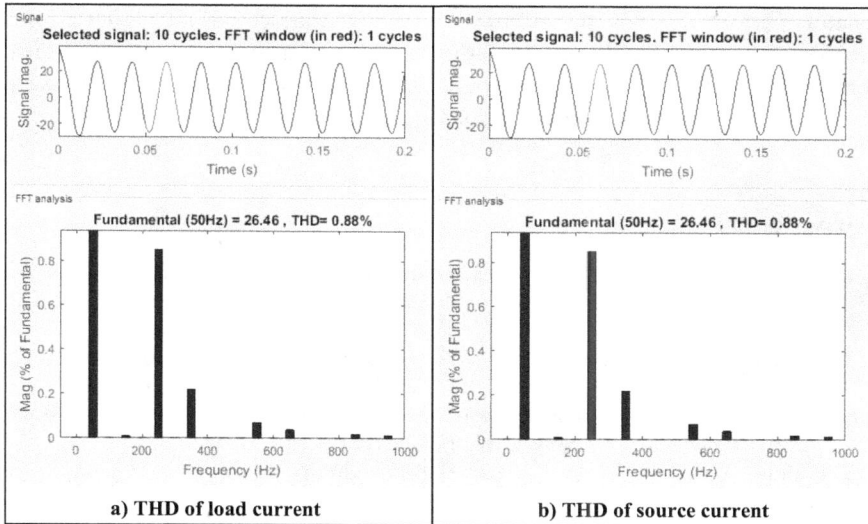

a) THD of load current   b) THD of source current

Figure 3.22   THD Analysis.

Table 3.3   Comparative Analysis THD of with & without PV-UPQC for Load and Source Voltage

THD% Load Voltage						
Conditions	*Without UPQC*			*With UPQC*		
Phase Current	Ia	Ib	Ic	Ia	Ib	Ic
Case 1	0.051	0.051	0.054	0.051	0.051	0.054
Case 2	3.32	3.32	3.32	1.81	1.81	1.81
Case 3	1.11	1.11	1.11	0.05	0.05	0.05
THD% Source Voltage						
	Without UPQC			*With UPQC*		
Case 1	0.052	0.052	0.052	0.052	0.052	0.052
Case 2	0.051	0.051	0.051	0.051	0.051	0.051
Case 3	1.13	1.13	1.13	0.05	0.05	0.05

dynamic performances of the system were evaluated under various modes of operation of grid voltage conditions, including sags, unbalances, and harmonics. Apart from series compensation, suppression of load harmonic currents carried out, such that an effective unified power conditioning was achieved.

References

1. S. Devassy, B. Singh, "Design and performance analysis of three-phase solar PV integrated UPQC," *IEEE Transactions on Industry Applications*, vol. 54, no. 1, pp. 73–81, 2017.
2. Leonardo Bruno Garcia Campanhol, S´ergio Augusto Oliveira da Silva, Azauri Albano de Oliveira Jr., Vin´ıcius D´ario Bacon, "Single-stage three-phase grid-tied PV system with universal filtering capability applied to DG systems and AC microgrids," *IEEE Transactions on Power Electronics*, vol. 32, no. 12, December 2017.
3. Rodrigo Augusto Modesto, S´ergio Augusto Oliveira da Silva, "A versatile unified power quality conditioner applied to three-phase four-wire distribution systems using a dual control strategy," *IEEE Transactions on Power Electronics*, vol. 31, no. 8, August 2016.
4. Guangqian Ding, Feng Gao, Hao Tian Cong Ma, Mengxing Chen, Guoqing He, Yingliang Liu's, "Adaptive DC-link voltage control of two-stage photovoltaic inverter during low voltage ride-through operation," *IEEE Transactions on Power Electronics*, vol. 31, no. 6, June 2016.
5. R. J. M. dos Santos, J. C. da Cunha, M. Mezaroba,"A simplified control technique for a dual unified power quality conditioner," *IEEE Transactions on Industrial Electronics*, vol. 61, no. 11, pp. 5851–5860, 2014.
6. V. Khadkikar, "Enhancing electric power quality using UPQC: A comprehensive overview," *IEEE transactions on Power Electronics*, vol. 27, no. 5, pp. 2284–2297, 2011.
7. H. Xiao, S. Xie, "Transformer less split-inductor neutral point clamped three-level PV grid-connected inverter," *IEEE Transactions on Power Electronics*, vol. 27, no. 4, pp. 1799–1808, April 2012.
8. Liming Liu, Hui Li, Yaosuo Xue, Wenxin Liu, "Decoupled active and reactive power control for large-scale grid-connected photovoltaic systems using cascaded modular multilevel converters," *IEEE Transactions on Power Electronics*, vol. 30, no. 1, January 2015.
9. M. C. Cavalcanti, A. M. Farias, K. C. Oliveira, F. A. S. Neves, J. L. Afonso, "Eliminating leakage currents in neutral point clamped inverters for photovoltaic systems," *IEEE Transactions on Industrial Electronics*, vol. 59, no. 1, pp. 435–443, January 2012.
10. B. H. Yong, V. K. Ramachandara Murthy, "Double tuned filter design for harmonic mitigation in grid connected solar PV," Power and Energy (PECon), December 2014.
11. Y. Tang, W. Yao, P. C. Loh, F. Blaabjerg, "Highly reliable transformer less photovoltaic inverters with leakage current and pulsating power elimination," *IEEE Transactions on Industrial Electronics*, vol. 63, no. 2, pp. 1016–1026, February 2016.
12. Y. Kim, H. Cha, B. M. Song, K. Y. Lee, "Design and control of a grid-connected three-phase 3-level NPC inverter for building integrated photovoltaic systems," *Proceedings IEEE PES Innovative Smart Grid Technology*, 2012, pp. 1–7.

13. M. Abolhassani,"Modular multipulse rectifier transformers in symmetrical cascaded H-bridge medium voltage drive," *IEEE Transactions Power Electronics*, vol. 27, no. 2, pp. 698–705, February 2012.

14. R. Torquato, W. Freitas, G. R. T. Hax, A. R. Donadon, R. Moya, "High frequency harmonic distortions measured in a Brazilian solar farm," 2016 17th International Conference on Harmonics and Quality of Power (ICHQP), pp. 623–627, 2016.

15. G. Ding, "Adaptive DC-link voltage control of two-stage photovoltaic inverter during low voltage ride-through operation," *IEEE Transactions on Power Electronics*, vol. 31, no. 6, pp. 4182–4194, June 2016.

16. Mohamed J. M. A. Rasul, H. V. Khang, Mohan Kolhe, "Harmonic mitigation of a grid-connected photovoltaic system using shunt active Filter," International Conference on Electrical Machines and Systems (ICEMS), 2017.

17. S. Bacha, D. Picault, B. Burger, I. Etxeberria-Otadui, J. Martins, "Photo voltaics in microgrids: An overview of grid integration and energy management aspects," *IEEE Industrial Electronics Magazine*, vol. 9, no. 1, pp. 33–46, March 2015.

18. Farhat Maissa and Sbita Lassâad Efficiency Boosting for PV Systems- MPPT Intelligent Control Based, Energy Efficiency Improvements in Smart Grid Components: Energy improvements in smart grid component. ISBN: 978-953-51-2038-4(2015).

19. Liming Liu, Hui Li, Yaosuo Xue, Wenxin Liu, "Decoupled active and reactive power control for large-scale grid-connected photovoltaic systems using cascaded modular multilevel converters," *IEEE Transactions on Power Electronics*, vol. 30, no. 1, January 2015.

20. Ine Vandoorn, Bart Meersman, Jeroen De Kooning, Lieven Vandevelde, "Controllable harmonic current sharing in islanded microgrids: DG units with programmable resistive behavior toward harmonics," *IEEE Transactions on Power Delivery*, vol.27, no. 2, April 2012.

21. Seddik Bacha, Damien Picault, Bruno Burger, Ion Etxeberria-Otadui, João Martins, "Photovoltaics in microgrids: An overview of gridintegration and energy management aspects," IEEE 1932–4529/15©2015IEEE Industrial Electronics Magazine.

22. S. A. O. Silva, L. P. Sampaio, F. M. Oliveira, F. R. Durand, "Feedforward DC-bus control loop applied to a single-phase grid-connected PV system operating with PSO-based MPPT technique and active power-line conditioning," *IET Renewable Power Generation*, vol. 11, no. 1, pp. 183–193, January 2017.

23. A. Muni Sankar, T. Devaraju, M. Vijaya Kumar, "Mitigation of power quality issues in grid integrated PV and wind energy systems using UPQC," *International Journal of Electrical and Electronics Engineering Research (IJEEER)*, vol. 4, no. 4, pp. 91–102, August 2014. ISSN(P): 2250–155X; ISSN(E): 2278–943X.

24. R. K. Varma, M. Salama, "Large-scale photovoltaic solar power integration in transmission and distribution networks," 2011 IEEE Power and Energy Society General Meeting, pp. 1–4, 2011.

25. H. Hu, Q. Shi, Z. He, J. He, S. Gao, "Potential harmonic resonance impacts of PV inverter filters on distribution systems," *IEEE Transactions Sustainable Energy*, vol. 6, no. 1, pp. 151–161, 2015.

26. O. S. Nduka, B. C. Pal, "Harmonic domain modeling of PV system for the assessment of grid integration impact," *IEEE Transactions Sustainable Energy*, vol. 8, no. 3, pp. 1154–1165, 2017.

27. Davood Yazdani, Alireza Bakhshai, Geza Joos, M. Mojiri, "A nonlinear adaptive synchronization technique for grid-connected distributed energy sources," *IEEE Transactions on Power Electronics*, vol. 23, no. 4, July 2008.

28. Davood Yazdani, Alireza Bakhshai, Geza Joos, M. Mojiri, "A nonlinear adaptive synchronization technique for grid-connected distributed energy sources," *IEEE Transactions on Power Electronics*, vol. 23, no. 4, July 2008.

29. B. H. Yong, V. K. Ramachandara Murthy, "Double tuned filter design for harmonic mitigation in grid connected solar PV," Power and Energy (PECon), December 2014.

30. P. Denholm, R. Margolis, "Very large-scale deployment of grid-connected solar photo Voltaics in the United States: Challenges and opportunities," in Proceedings National Renewable Energy Laboratory Conference Paper Preprint Solar US Department of Energy, 2006.

31. H. Kanchev, D. Lu, F. Colas, V. Lazarov, B. Francois, "Energy management and operational planning of a microgrid with a PV-based active generator for smart grid applications," *IEEE Transactions on Industrial Electronics*, vol. 58, no. 10, pp. 4583–4593, October 2011.

32. P. C. Loh, D. Li, Y. K. Chai, F. Blaabjerg, "Autonomous control of interlinking converter with energy storage in hybrid AC–DC microgrid," *IEEE Transactions on Industry Applications*, vol. 49, no. 3, pp. 1374–1383, May 2013.

33. H. Beltran, I. Etxeberria-Otadui, E. Belenguer, P. Rodriguez, "Power management strategies and energy storage needs to increase the operability of photovoltaic plants," *Journal of Renewable and Sustainable Energy*, vol. 4, 063101, 2012.

34. D. Picault, B. Raison, S. Bacha, J. De La Casa, J. Aguilera, "Forecasting photovoltaic array power production subject to mismatch losses," *Solar Energy*, vol. 84, no. 7, pp. 1301–1309, 2010.

35. B. H. Chowdhury, "Central-station photovoltaic plant with energy storage for utility peak load leveling," Proceedings IEEE 24th Intersociety Energy Conversion Engineering Conference, IECEC-89, August 1989, pp. 731–736.

36. L. Liu, H. Li, Y. Xue, W. Liu, "Reactive power compensation and optimization strategy for grid-interactive cascaded photovoltaic systems," *IEEE Transactions on Power Electronics*, vol. 30, no. 1, pp. 188–202, January 2015.

37. Liming Liu, Hui Li, Yaosuo Xue, Wenxin Liu, "Decoupled active and reactive power control for large-scale grid-connected photovoltaic systems using cascaded modular multilevel converters," *IEEE Transactions on Power Electronics*, vol. 30, no. 1, January 2015.

Chapter 4

A Hybrid Energy Management for Stand-Alone Microgrids Using Grey Wolf Optimization System

Siddharth Shukla and Amit Gupta

Contents

DOI: 10.1201/9781003301820-4

4.1 Introduction

Since conventional energy sources are having a lot of issues that are having a bigger influence on the environment, renewable energy sources (RESs) are vital. RESs are better able to reduce pollution and global warming [1]. Because of advancements in technology and environmental concerns, RES-based distributed generation (DG) is becoming more significant [2]. Numerous optimization strategies have been developed to regulate various power electronic devices for improved flexibility while integrating various alternative sources like solar, wind, battery, etc. that are crucial in meeting the vast energy needs [3, 4]. Due to the limits of conventional energy sources, DG power generation is more significant since it meets end users' ongoing demands for energy while providing more productivity and better maximum quality [5]. Power generation has become a difficult undertaking because of the rise in population and energy consumption. Hence, RESs have taken over since conventional sources alone are unable to meet the demands.

As a result of the increase in energy consumption, reducing greenhouse gas (GHG) emissions, increasing energy efficiency, and having an abundance of clean electricity have all emerged as major issues in the energy sector. Due to their accessibility, scalability, dependability, enhanced power quality, greater flexibility, cheaper cost, and reduced environmental effect, alternative energy sources like wind, biomass, solar, and hydro are growing in popularity, especially in rural areas and remote islands [6]. However, these intrinsic shortcomings of DGs, including as intermittency and the ensuing power quality problems, become more noticeable as a result of their substantial penetration into the power grid [7].

Microgrids are a new technology that can be used to address these problems locally and guarantee a steady supply of electricity. Microgrids can be either grid-connected microgrids, which are connected to the conventional utility grid, or self-contained power grids (stand-alone microgrids), which rely on local resources. Regardless of their design, microgrids have proved effective at lowering energy costs and CO_2 emissions [8]. However, the microgrid has considerable challenges as a result of the limitations of renewable energy sources, particularly in terms of stability and power quality. Energy storage has therefore been recognized as a potential strategy for addressing these issues while also enhancing system security and flexibility. ESSs can be used to store extra power during periods of high availability and then release it when the grid is experiencing a power shortage. ESSs can also participate in the market by buying energy from the upstream grid during off-peak hours and selling it back to the upstream grid during high demand hours, thanks to the electrical market's real-time pricing. Additionally, additional grid services could be offered by microgrid storage devices to enhance power quality [9].

Two often used sources of energy are solar and wind [10]. The efficiency of RES will increase with the use of power electronic devices in the power system. To meet the required load demand and enhance the quality of the power for customers, one or more HRES system resources are incorporated into the distribution network [8]. The interaction of these two variables may result in more PQ issues since the system oscillates and exhibits features like sags, swells, interrupted harmonics, etc. This causes constant voltage fluctuations and constant tripping due to variance [11].

Due to the variability and erratic nature of renewable energy sources like wind turbines (WTs) and photovoltaic (PV) modules, the usage of storage devices in microgrids has become crucial [7]. These storage systems can store excess renewable energy during off-peak load demands or contribute extra power to the grid in the event of a power loss. The involvement (charging/discharging) of storage devices in microgrids is essential for maintaining power balance, as was previously mentioned. However, a very little battery capacity would result in insufficient power, which would lead to instability or increase the cost of using conventional fuel. An excessive battery capacity would increase cost. As a result, determining the optimal storage device capacity or size is critical for avoiding microgrid dispatch issues and optimizing operation costs [12].

In this chapter, a hybrid system is considered that combines PV, wind, and battery energy storage system (BESS) with the distribution system to meet the necessary load demand. Hybrid DGs are an issue because they cause instability. Most of the system's problems are related to power quality, such as harmonics and sag problems. These PQ issues particularly affect HRES systems because of the intermittent nature of solar and wind energy sources and their influence on environmental conditions. The flexible AC transmission systems (FACTS) devices and filter can help to mitigate these power quality difficulties in the system. Thanks to advancements in power electronics device and control approaches, many devices are now available and classified as series devices and shunt devices. The dynamic voltage restorer (DVR) and the static synchronous series compensator (SSSC) is the device used to maintain the voltage. To inject reactive and harmonic components in current mode and control the load voltage in voltage control mode, respectively, static compensators (STAT-COMs), distribution STATCOM (DSTATCOM), and thyristor-controlled reactors (TCR) are utilized [13–18]. To maintain stability and minimize power quality problems, HRES-based DGs must employ an appropriate controlling method.

Additionally, the smart energy-management systems seek to determine the ideal battery capacity while also lowering the amount of conventional fuel used and total operating expenses. The ideal size for renewable energy sources has thus been the subject of numerous studies. The GWO results demonstrate that, in terms of both solution quality and computational effectiveness, the suggested strategy can locate the best global optima in the optimization problem. The following is a summary of this study's primary contribution:

1. It offers a new approach for intelligent energy management that expands the use of renewable energy sources and lessens the microgrid's reliance on fossil fuels.
2. It considers the impact of the ideal battery-storage system size on the management of operations as well as the total cost of the microgrid.

4.2 Grey Wolf Optimizer

One of the potent meta-heuristic algorithms proposed by [19] is the grey wolf optimizer (GWO). One of the unconventional algorithms, GWO exhibits higher performance while handling a variety of challenging situations. It can compete with algorithms like PSO, GA, and many more. Grey wolves served as the GWO's inspiration. The Canidae family, which includes members of the grey wolf family, is the apex predator in the food chain. These wolves live in packs of five to 12 individuals.

The alpha, often known as the pack leader, is in charge of the wolf pack. The pack's second level after the alpha, beta, enforces the alpha's directives and offers feedback to the alpha. The omega, or lowest rank in the grey wolf hierarchy, frequently serves as the scapegoat. A lobe is referred to as a delta if it is not an alpha, beta, or omega lobe. Typically, the delta performs the roles of scouts, sentries, elders, hunters, and caregivers. The social dominance structure of grey wolves is shown in Figure 4.1. In the following subsection, the processes of the GWO mathematical formulation for hunting the prey are explained.

4.3 Mathematical Formulation of the Grey Wolf Optimizer

This GWO, which is depicted in Figure 4.2, is a novel metaheuristic method that relies on swarm intelligence and is inspired by the thought of grey wolves when they are pursuing prey [20]. They remain as a pack and are placed in a position to conduct the hunting process. The mathematical approach is designed by giving the group the best fittest answer, which is then followed by the groups. To start the hunt for the prey, they seal off a corridor around the injured and provide calculations (1).

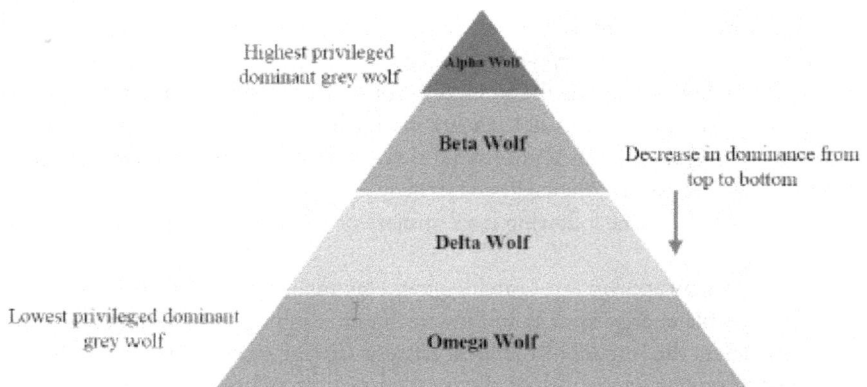

Figure 4.1 The Dominance Hierarchy of Grey Wolves.

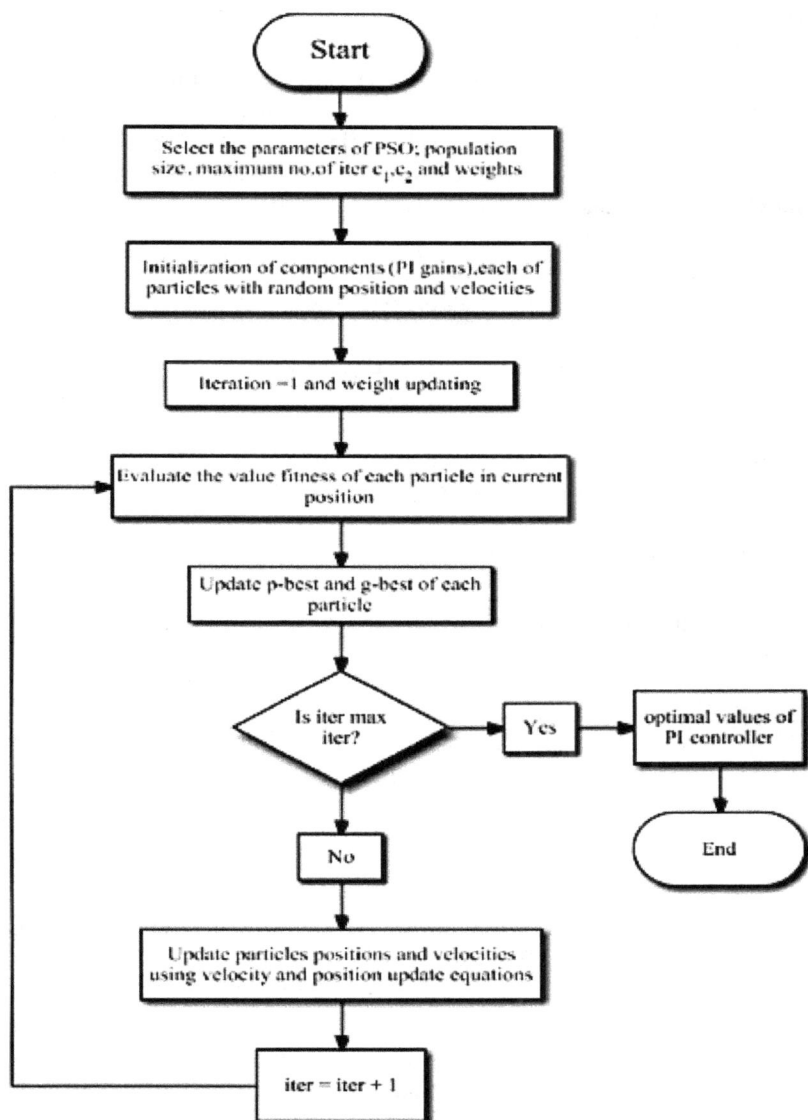

Figure 4.2   Flow Chart of GWO.

The alpha (α) wolf position is supposed to be the best answer in the suggested GWO algorithm, with beta (β) and delta (δ)) being the second and third best replies, respectively. The behaviors of the GWO can be described mathematically in this way. The remaining optimization issue solutions are represented by omega (ω). The hunting in the GWO algorithm is directed by α, β, and δ, while the omega follows these wolves. It is not an easy task for the grey wolf pack to hunt prey since there are various steps that must be followed carefully to produce the best hunting results. These steps include chasing, tracking, surrounding, and updating the wolves' positions in accordance with the dam. Grey wolves typically circle their prey to stop it from moving after the chase and tracking stages, and the following set of equations quantitatively illustrates this behavior:

$$D \rightarrow = |C \rightarrow .Xp \rightarrow (t) - X \rightarrow (t)| \tag{4.1}$$

$$X \rightarrow (t+1) = Xp \rightarrow (t) - A \rightarrow .D \rightarrow \tag{4.2}$$

where,
$A\rightarrow$ and $D\rightarrow$ are the vector coefficients,
$Xp\rightarrow$ is the vector location of the prey,
t is the current iteration,
$X\rightarrow$ is the location vector of a grey wolf.

The encircling equations can be obtained by finding the coefficients vectors $A\rightarrow$ and $C\rightarrow$ as follows:

$$A \rightarrow = 2a \rightarrow .r1 \rightarrow -a \rightarrow \tag{4.3}$$

$$C \rightarrow = 2.r2 \tag{4.4}$$

$$a = 2 - t * 2\text{maximum iteration} \tag{4.5}$$

where,
$a\rightarrow$ is linearly changed from 2 to 0 during the algorithm iterations, and r1 and r2 are random values between (0, 1).

The top ideas from each iteration are kept, and the other wolves (omega) adjust their places in accordance with those iterations. The following sets of equations can be used to express the GWO algorithm's updating processes:

$$D_{Alpha} \rightarrow = |C1 \rightarrow .XAlpha \rightarrow -X \rightarrow| \tag{4.6}$$

$$D_{Beta} \rightarrow = |C2 \rightarrow .XBeta \rightarrow -X \rightarrow| \tag{4.7}$$

$$D_{Delta} \rightarrow = |C3 \rightarrow .XDelta \rightarrow -X \rightarrow| \tag{4.8}$$

The vector positions of the prey can be determined based on the alpha, beta, and delta positions using the following equations:

$$X1 \to = | \, xAlpha \to -A1 \to .DAlpha \to | \tag{4.9}$$

$$X2 \to = | \, xBeta \to -A2 \to .DBeta \to | \tag{4.10}$$

$$X3 \to = | \, XDelta \to -A3 \to .DDelta \to | \tag{4.11}$$

$$X \to (t+1) = X1 \to + X2 \to + X3 \to 3 \tag{4.12}$$

The exploration and exploitation of the grey wolf agents based on the parameter A, if the parameter A is ($|A| \geq 1$), then half of the iterations are devoted to exploration. Meanwhile, when ($|A| < 1$) the other half of iterations are devoted to exploitation. The pseudo-code of the GWO algorithm is presented in the following form.

Algorithm 4.1: GWO Pseudo-Code

Initialize the locations of the grey wolf population Xi (I = 1, 2, . . . , n
Initialize a, A, and C
Calculate the objective function value for each grey wolf agent
Set: $X\alpha$ as best result of the search agents
$X\beta$ as the second-best result of the search agents
$X\delta$ as the third-best result of the search agents
While (t < max number of iteration) the termination criterion is not satisfied **do**
Initialize $r1$ and $r2$ values
Update a by Equation (5)
Update A by Equation (3)
Update C by Equation (4)
For i
For j
Update the positions of each grey wolf agent by using Equations (6)–(12)
End j
End i
Calculate the fitness of all agents with the new positions
$T = t + 1$
End while
return $X\alpha$

4.4 The GWO Implementation of the Optimal Operation Management of the Microgrid

By identifying the appropriate values for the parameters that help to reduce the operating cost of the generating sources in the microgrid and satisfy all the constraints in each phase of the GWO algorithm, the GWO is employed in this case to solve the operational management challenges in the microgrid. The performance flowchart of the grey wolf algorithm for microgrid operation management is shown in Figure 4.3.

However, due to the limitations of the generation sources and the energy-management constraints, the function handle (Algorithm 4.2) should be utilized as follows:

Algorithm 4.2: Function Handle

For $t = 1$ to NT **do**
For $m = 1$ to NOA **do**
Part 1: power balance and generation source capacity handling
Calculate the power difference between the generation sources and load demand
$Pdiff = (P\_MT,t\, u\_MT,t + P\_FC,t\, u\_FC,t + P\_PV,t + P\_WT,t + P\_BES,t\, u\_BES,t + P\_grid,) - PD,t$;
Select random generation sources based on their capacity
While $P\_diff \neq 0$ **do**
Subtract P_diffm,t from the selected units
Check the capabilities of the generation units based on lower and upper limits as follows:
If $Pm,FC,t < PFC,min$ **then** $Pm,FC, = PFC,min$;
or $Pm,MT,t < PMT,min$ **then** $Pm,MT,t = PMT,min$;
or $Pm,grid,t < Pgrid,min$ **then** $Pm,grid,t = Pgrid,min$;
or $Pm,BES,t < PBES,min$ **then** $Pm,BES,t = PBES,min$;
Elseif $Pm,FC,t > PFC,max$ **then** $Pm,FC, = PFC,max$;
or $Pm,MT,t > PMT,max$ **then** $Pm,MT,t = PMT,max$;
or $Pm,grid,t > Pgrid,max$ **then** $Pm,grid,t = Pgrid,max$;
or $Pm,BES,t < PBES,max$ **then** $Pm,BES,t = PBES,max$;
End if
Calculate P_diffm,t
Select another generation units randomly
End while
Part 2: ORs handling
Calculate objective function (*ft*) by using Equation (2)
If $P\_MT,t\, u\_MT,t + P\_FC,t\, u\_FC,t + P\_PV,t + P\_WT,t + P\_BES,t\, u\_BES,t + P\_grid,t < PD,t + ORt$
Then $ft = ft +$ Penalty_Factor $\times$ ($P\_MT,t\, u\_MT,t + P\_FC,t\, u\_FC,t + P\_PV,t + P\_WT,t + P\_BES,t\, u\_BES,t + P\_grid,t - (PD,t + ORt$))
End if
End for m
Calculate Equation (1)
End for t

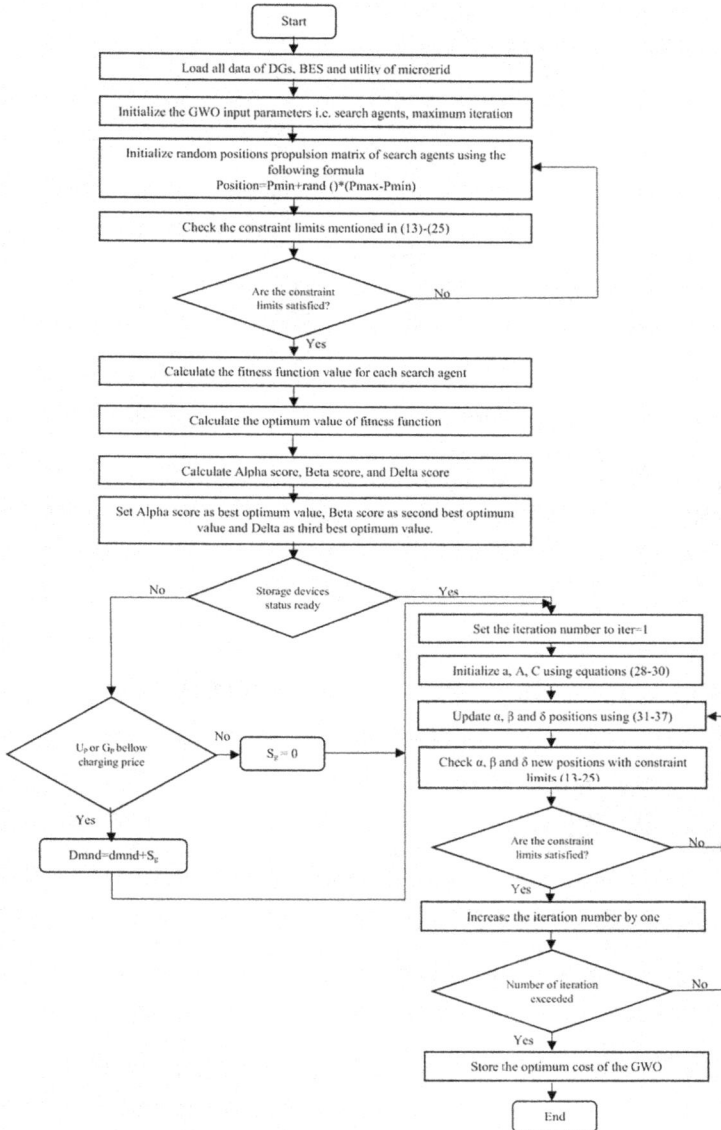

Figure 4.3 The Grey Wolf Optimization Algorithm Flowchart Used in the Microgrid System.

Since the OR's constraints should be met, the penalty factor value considered in this chapter is 10.

The grey wolves' numbers (search agents) and the iterations numbers are set. The population vector of the GWO can be represented as follows:

$$X = \begin{bmatrix} x_1^1 & \cdots & x_n^1 \\ \vdots & \ddots & \vdots \\ x_1^P & \cdots & x_n^P \end{bmatrix} \qquad (4.13)$$

where n represents the control variable numbers or search agents' positions. The population number (grey wolves) is represented by p.

We used a monitoring technique for the microgrid operations to determine the charging and discharging rules for storage devices that integrated with the grey wolf algorithm to reduce the operational cost in the microgrid. The approach is focused on determining the microgrid's economical pricing factor within 24 hours. As a result, the decision to charge or discharge a battery in a microgrid is based on the difference between the highest and lowest mean prices of the utility market and the DGs over the preceding 24 hours. By contrasting the dispatch cost of the storage devices with other DGs like gas and the utility grid's dynamic generation price, the grey wolf algorithm decides how much to charge the storage devices.

4.5 Proposed HRES (Hybrid Renewable Energy Sources System with GWO)

Due to the rapid industrial development, there are now more distribution organizations to meet the demand for energy. The essential demand for electricity cannot be satisfied by the amount of electricity generated by fossil sources. These conventional sources have a negative impact on the environment and accelerate global warming. Renewable energy sources are being pushed to supply the energy required to fulfill load demand in order to address these problems.

Solar and wind energy, as well as other renewable energy sources, are extensively adopted by the distribution system. When PV and WT are coupled, it is believed that HRES is the best choice for increased system dependability and efficiency. Even though there are problems with the flexibility and stability of the power quality, the distribution system's HRES can manage the consumers' necessary load requirements. These PQ issues need to be avoided for the system to function dependably [20]. FACT devices were developed to help with power quality problems including harmonics, sag, and swell as power electronics advanced. In this work, voltage sag, current sag, and THD problems in power systems are addressed using UPQC.

The proposed HRES system interfaces the PV and WT combination with the grid. BESS is designed to meet load demand under harsh environmental conditions because the sources considered are intermittent. The unexpected loads and

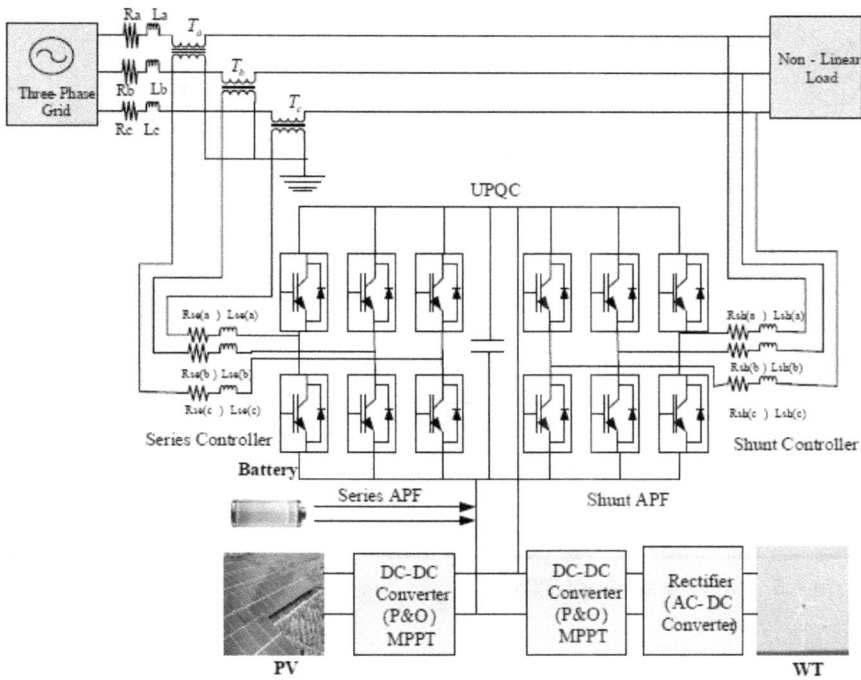

Figure 4.4   Block Diagram of Grid-Connected HRES.

non-linear loads cause PQ issues in the HRES system [21]. These problems lead to voltage instability and reactive power imbalances. This chapter [22] discusses PQ concerns such voltage sag, current sag, actual power, reactive power, and THDs. The HRES system was developed using UPQC to improve voltage regulation using a FOPID controller in order to solve these issues. The system is managed using optimization methods like GWO. In this chapter, UPQC is used to reduce issues with voltage sag, current sag, and THD in power systems.

The proposed system block design, shown in Figure 4.4, is primarily made up of distributed generation employing a PV-Wind combination coupled to the inverter to interface with the grid via DC-link capacitor [22, 23]. In the event of a system collapse, BESS is designed to store energy and supplies. Utilizing perturb and observe (P&O) MPPT, solar and wind energy is collected. GWO generates a reference DC link voltage. The reference DC link voltage is set to its default value when there is no solar energy present.

The power quality issues of the HRES, which are mostly brought on by grid-side faults, non-linear loads, and sudden loads, are considered [23]. The proposed system is constructed using UPQC (Figure 4.5), which employs control techniques with the help of series and shunt controllers to account for the errors and guarantee

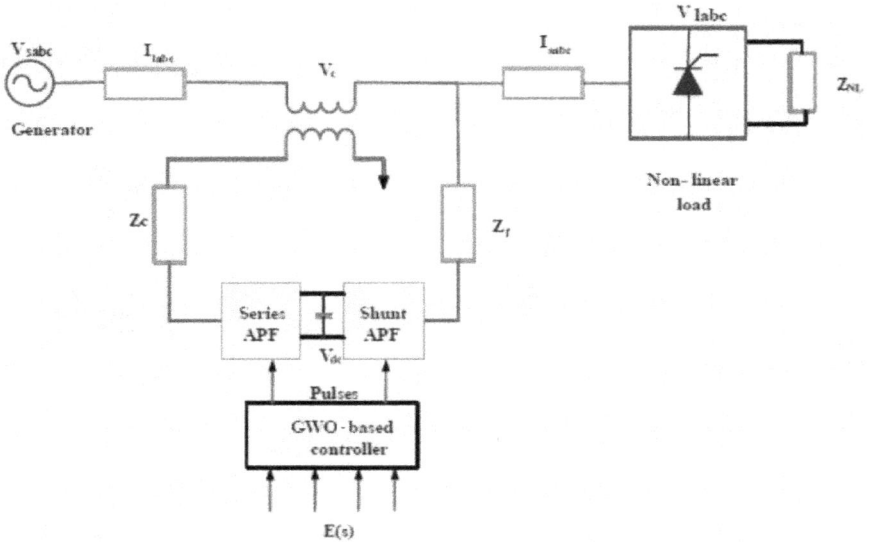

Figure 4.5   Architecture of UPQC.

stable operation. Power compensation in sag conditions is possible by providing the best gain parameters to the FOPID controller, which filters and injects the required power. To conduct the control operation under PQ conditions, the GWO, which is utilized to operate the FOPID controller, is supposed to select the relevant values [24].

4.6 FOPID Controller

Figure 4.6 shows the FOPID's basic diagram. The control output u is produced with the aid of the error signal e (s) (s). Voltage and current variations in the HRES system are a source of poor power quality, which is why FOPID with GWO optimization is used. The mathematical formulation of the FOPID controller's control signal is as follows:

$$u(s) = Kp + Ki\ D^{-\lambda}e(s) + Kd\ D^{\mu}e(s)$$

The following steps are required during the design of the controller:

1. KP is controlled to reduce steady-state error and rising time.
2. Kd is controlled to reduce settling time and overshoot.
3. Ki is controlled to get eliminate of steady-state error.
4. $D^{-\lambda}$ and D^{μ} are fractional order parameters.

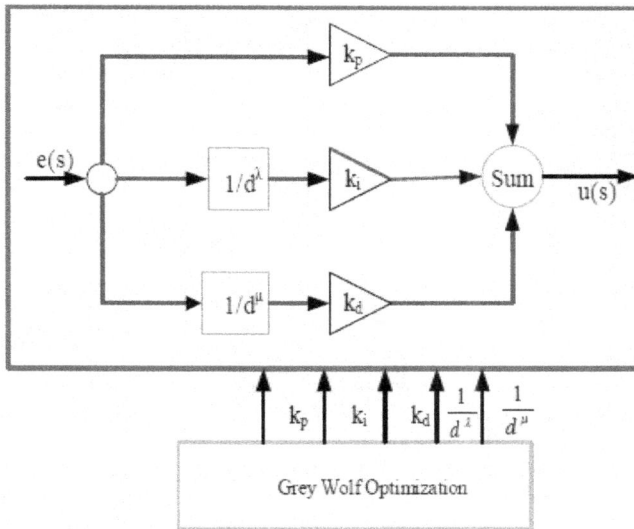

Figure 4.6   Fractional Order Proportional Integral Derivative Controller.

Sag faults are produced as a result of the placement of non-linear loads in the proposed HRES grid-connected systems. The proposed systems are built with a FOPID–GWO-based controlling technique to reduce and balance load demand. These systems fix the speed and irradiance of the PV and WT to a fixed value. PV emits 1000W/m2 of radiation, while the average wind speed is 12m/s. This hybrid solution is intended to handle the load requirement while reducing PQ issues.

4.7 Conclusion

To increase the power quality in a grid-connected hybrid system using battery storage without interfering with the regular operation of real power transfer, a GW optimization is implemented as a controller for the distributed power system. The GWO is used to construct the ideal set of parameters, and the PI approach is used to forecast the ideal control signals. When compared to other methods, the GWO-FOPID controller produces the best results in terms of THDs by lowering the system's harmonic levels.

References

1. M. Bajaj and A. K. Singh, "Grid integrated renewable DG systems: A review of power quality challenges and state-of-the-art mitigation techniques", *International Journal of Energy Research*, Vol. 44, No. 1, pp. 26–69, 2020.

2. W. U. Tareen, S. Mekhilef, M. Seyedmahmoudian, and B. Horan, "Active power filter (APF) for mitigation of power quality issues in grid integration of wind and photovoltaic energy conversion system", *Renewable and Sustainable Energy Reviews*, Vol. 70, pp. 635–655, 2017.
3. S. Paramanik, K. Sarker, D. Chatterjee, and S. K. Goswami, "Smart grid power quality improvement using modified UPQC", *Devices for Integrated Circuit (DevIC)*, Kalyani, India, pp. 356–360, 2019, doi: 10.1109/DEVIC.2019.8783704.
4. R. Sedaghati and M. R. Shakarami, "A novel control strategy and power management of hybrid PV/FC/SC/battery renewable power system-based grid connected microgrid", *Sustainable Cities and Society*, Vol. 44, pp. 830–843, 2019.
5. N. B. Kumar, K. K. Chowdary, and V. Subrahmanyam, "Renewable energy hybrid power system with improvement of power quality in grid by using DVSI", *Renewable Energy*, Vol. 6, No. 09, 2019.
6. B. Ye, K. Zhang, J. Jiang, L. Miao, and J. Li, "Towards a 90% renewable energy future: A case study of an island in the South China Sea," *Energy Conversion and Management*, Vol. 142, pp. 28–41, 2017/06/15/.
7. A. A. Moghaddam, A. Seifi, T. Niknam, and M. R. A. Pahlavani, "Multi-objective operation management of a renewable MG (micro-grid) with back-up micro-turbine/fuel cell/battery hybrid power source", *Energy*, Vol. 36, pp. 6490–6507, 2011.
8. H. Kuang, S. Li, and Z. Wu, "Discussion on advantages and disadvantages of distributed generation connected to the grid", Electrical and Control Engineering (ICECE), 2011 International Conference on, 2011, pp. 170–173.
9. G. Mallesham and C. S. Kumar, "Power quality improvement of weak hybrid pemfc and scig grid using upqc", In *Advances in Decision Sciences, Image Processing, Security and Computer Vision*, pp. 406–413, Springer, Cham, 2020.
10. Q. Liu, Y. Li, S. Hu and L. Luo, "Power quality improvement using controllable inductive power filtering method for industrial DC supply system", *Control Engineering Practice*, Vol. 83, pp. 1–10, 2019.
11. S. Ghosh and M. H. Ali, "Minimisation of adverse effects of time delay on power quality enhancement in hybrid grid", *IEEE Systems Journal*, Vol. 13, No. 3, pp. 3091–3101, 2019.
12. J. Hussain, M. Hussain, S. Raza, and M. Siddique, "Power quality improvement of grid connected wind energy system using DSTATCOMBESS", *International Journal of Renewable Energy Research (IJRER)*, Vol. 9, No. 3, pp. 1388–1397, 2019.
13. F. Nejabatkhah, Y. W. Li, and H. Tian, "Power quality control of smart hybrid AC/DC microgrids: An overview", *IEEE Access*, Vol. 7, pp. 52295–52318, 2019.
14. A. Bhargava and S. Verma, "Power quality enhancement using unified power flow controller in grid connected hybrid PV/Wind system", 2019 International Conference on Communication and Electronics Systems (ICCES), pp. 2064–2069, IEEE, 2019.
15. B. S. Goud, B. L. Rao, and C. R. Reddy, "Essentials for grid integration of hybrid renewable energy systems: A brief review", *International Journal of Renewable Energy Research (IJRER)*, Vol. 10, No. 2, pp. 813–830, 2020.
16. B. S. Goud and B. L. Rao, "An intelligent technique for optimal power quality enhancement (OPQE) in an HRES grid-connected system: ESA technique", *International Journal of Renewable Energy Research (IJRER)*, Vol. 10, No. 1, pp. 317–328, 2020.
17. G. Meerimatha and B. L. Rao, "Novel reconfiguration approach to reduce line losses of the photovoltaic array under various shading conditions", *Energy*, Vol. 196, p. 117120, 2020.
18. B. S. Goud and B. L. Rao, "Review of optimisation techniques for integrated hybrid distribution generation", *International Journal of Innovative Technology and Exploring Engineering (IJITEE)*, Vol. 8, No. 5, pp. 527–533, 2019.

19. A. Billionnet, M. C. Costa, and P. L. Poirion, "Robust optimal sizing of a hybrid energy stand-alone system", *European Journal of Operational Research*, Vol. 254, pp. 565–575, 2016.
20. A. K. Mishra, S. R. Das, P. K. Ray, R. K. Mallick, A. Mohanty, and D. K. Mishra, "PSO-GWO optimized fractional order PID based hybrid shunt active power filter for power quality improvements", *IEEE Access*, Vol. 8, pp. 74497–74512, 2020.
21. M. Chiandone, C. Tam, R. Campaner, and G. Sulligoi, "Electrical storage in distribution grids with renewable energy sources," 2017 IEEE 6th International Conference on Renewable Energy Research and Applications (ICRERA), San Diego, CA, pp. 880–885, 2017, doi: 10.1109/ICRERA.2017.8191186.
22. C. R. Reddy and K. H. Reddy. "Islanding detection techniques for grid integrated DG–a review", *International Journal of Renewable Energy Research (IJRER)*, Vol. 9, No. 2, pp. 960–977, 2019.
23. R. Thumu and K. H Reddy, "PI, fuzzy based controllers for FACTS device in grid connected PV system", *International Journal of Integrated of Integrated Engineering*, Vol. 11, No. 6, pp. 176–185, 2019.
24. V. K. Kamboj, S. Bath, and J. Dhillon, "Solution of non-convex economic load dispatch problem using grey wolf optimizer", *Neural Computing & Applications*, Vol. 27, pp. 1301–1316, 2016.

Chapter 5

Energy Management of Nanogrid through Flair of Deep Learning from IoT Environments

Vandana Sondhiya, Kaustubh Dwivedi,
Shekh Kulsum Almas, and Nagendra Singh

Contents

DOI: 10.1201/9781003301820-5

5.1 Introduction

An intelligent algorithm-based IoT is competent in a holistic approach to handshake the world from every perspective. The goal of an IoT is to accomplish with veracity in the digital transformation through wireless communications such as smartphones, sensors, Bluetooth mesh networking, Light fidelity (Li-fi), radiofrequency waves, actuators, etc. in the ecosystem with a unique addressing mode to make synergy. Indeed, IoT is a physical object receiving signals from the sensor for coordination control in any ecosystem [1].

Due to computational proficiency and dynamic learning approach, it finds application in a variety of domains like surveillance, modern cities, homes, manufacturing, textiles, smart healthcare, smart grid, testing, monitoring, and so on. Consequently, the IoT paradigm is evolving in an integrated approach in the field of feature extraction, handling heterogeneous data, security systems, privacy, etc. However, the IoT paradigm encounters the formidable challenge of satisfying the requirements up to a utopia because of handling heterogeneous fields of applications [2].

Furthermore, with the increased use of IoT technologies, risk becomes prominent, especially in the areas of privacy and security. To overcome these issues certain measures including guidelines, regulatory framework, and comprehensive standards are taken [3]. Moreover, because of the intermittent behavior of the distributed generation, smart energy systems are associated with the existing nano-grid to enhance the grid resilience, peak hour savings, stability, efficiency, and of course more choices for the end users. For superior performance, an optimal energy management system will be the key enabler of the nanogrid. Furthermore, overestimation or underestimation of renewable energy generation will divert the optimization in an erroneous direction [4].

Major concerns in the centralized power system are carbon dioxide emissions in the environment, line losses, increasing energy demand, failure of equipment due to outages, blackout, depletion of fossil fuels, and so on. Another major issue is that most people live in the countryside or inaccessible areas, which do not have access to electricity, and extending the network is often considered complicated and uneconomical. Limitless energy resources are available such as biomass, geothermal, breeze, solar, hydro, wave, or tidal with huge potential for alternating energy generation [5, 6]. Moreover, the huge investment in the installation of shared power supplies and the intermittent nature of energy production by renewable resources is the major challenge to be widely accepted [7, 8]. To overcome these deficiencies and increase efficiency, researchers have provided a unique power distribution control system—microgrid [9, 10]. Microgrids combine the distributed generation units, wind turbines, storage, energy units, and control units. Nanogrid differs from the microgrid in terms of scale. A nanogrid is more compact, flexible, reliable, and operated in networked or islanded mode through a gateway. Nanogrid becomes more flexible with the gateway; it may be one-way or two-way for exchanges of power, interoperability, and seamless communications along with peripherals [11, 12]. The controller is the brain of the nanogrid to command the level of energy supplied to the loads, opt the local electricity cost/unit, and manage internal storage, peak shifting, etc. Nanogrid includes energy storage to explore stability and flexibility in the planning of the renewable energy resources requirement in the grid. Usually, sunlight is not available during the night and rainy season similarly; the wind is not available at the same speed all the time and varies with day and night. Hence, because of the periodic nature of renewable energy resources, a smart energy management system would be a better solution to forecast the demand profile and move the loads from crest periods to bottom periods [13–18]. Yet been few kinds of research have been conducted in the area of energy management. Hereafter, more investigations are to be concentrated on the energy management strategy of nanogrid. Network characteristics can be enhanced by adopting optimization techniques to utilize the resources in an optimum way, capacity, and power factor of DGs. Lots of approaches have been proposed to control and energy management of MG/NG [19–21]. Despite providing system interoperability and manageability of the power setup network, it has several vulnerabilities related to an only point of collapse and the limiting of reliability and adaptability. Heuristic research and significant attention must be envisaged to make it self-sufficient through intelligent algorithm with veracity [22–25].

Integration of control circuits with the peripheral units is a major concern in operating NGs. Reliability of centralized control is affected due to the use of a single controller. Apart from this, synergy of digital transformation and deployment of distributed generation will be a trustworthy and holistic approach especially designed for implementation in allocated systems of isolated area as NGs [26, 27].

To improve energy management system of battery energy storage (BES) and to optimize the renewable energy generation with veracity, distinguished machine learning algorithm has been deployed. Broadly, machine learning algorithm has been classified as supervised learning, unsupervised learning, and reinforcement learning. Typical machine learning algorithm set up explosion by utilizing various components such as data, model, parameters, learning algorithm such as perceptron learning algorithm, gradient descent, etc., objective/error/loss function to minimize losses [28]. Artificial neural network (ANN), multilayer perceptron (MLP), back propagation, gradient-descent method, recurrent neural network (RNN), decision tree, support vector machine (SVM), and so on are the tools of the artificial neural network. These learning methods use perceptron to eventually be able to learn, make decisions, and translate languages. A plethora of domains where machine learning, which is a family member of artificial intelligence, shows superior performance are in image processing, speech recognition, magnetic resonance imaging (MRI), smart packaging, smart grid, etc. Despite that, the scalability of these tools is inefficient in handling voluminous data due to slow convergence, over fitting, etc. [29].

In recent years, increased emphasis on handling enormous theoretical data and vanishing numerical instability substantial research has involved dynamic learning approach, that is, deep learning (DL) and deep reinforcement learning (DRL) in the field of artificial intelligence [30].

This chapter is categorized in 6 sections. Section 5.2 summarizes the process of deep learning, its advantages, and extensive applications. Section 5.3 introduces the features of nano-grid. Section 5.4 discusses the efficient energy management of nanogrid with the essence of deep learning.

Section 5.5 identifies the challenges and solutions of energy management of nano-grid. Section 5.6 compares machine learning with deep learning, and section 7 presents the conclusion and future scope.

5.2 Features of Deep Learning

This section introduces a holistic approach in discussing the energy management system of nano-grid or cluster of nano-grid.

5.2.1 Architecture of Deep Learning

Deep learning (DL) is a subfield of artificial intelligence.

A brief definition of the DL: "DL is a type of machine learning based on artificial neural network in which multiple (deep) layers of processing are used to extract proficient level features from data with veracity." Figure 5.1 shows the level of deep learning [29].

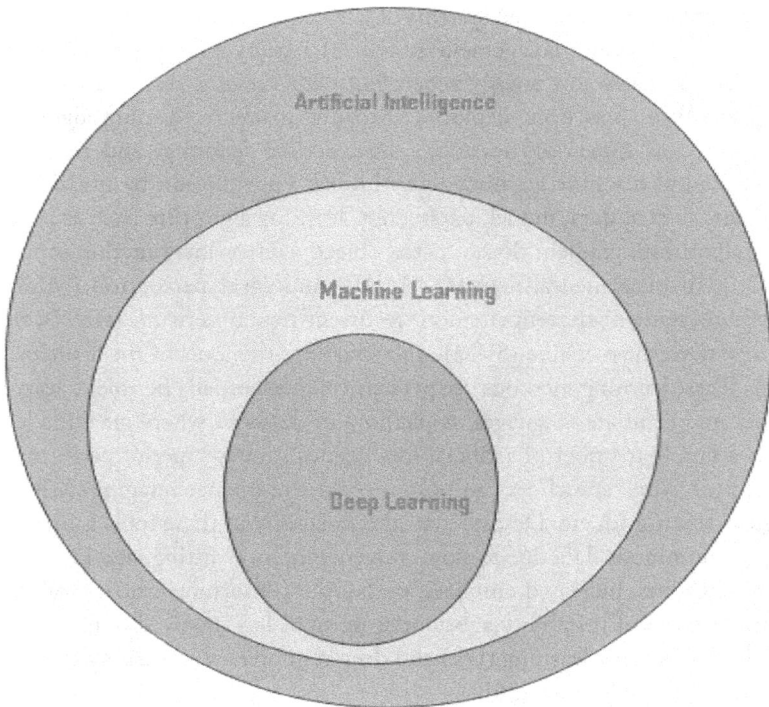

Figure 5.1   Level of Deep Learning [32].

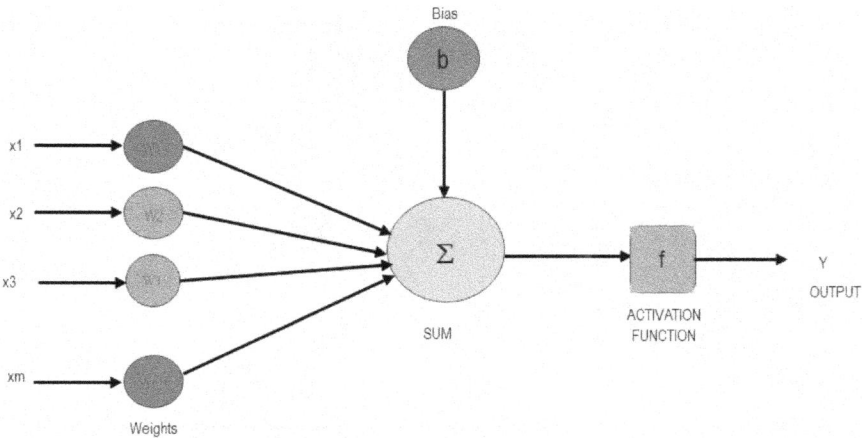

Figure 5.2   Deep Learning Neurons versus Biological Neurons [34].

The most fundamental unit of deep neural network is an artificial network. Motivation comes from biological neurons depicted in Figure 5.2, which is analogous to neural cells that resemble neural processing units [33].

Deep learning mimics the human brain as neurons connected through massive parallel neuron network, and each neuron has a distinguished role to play such as processing visual data, speed data, and so on to certain stimulus.

Deep architecture has a multilayer perceptron (MLP), which consists of an output layer envisaging a supervised target for an input layer, which is raw sensory units. Between these layers, hidden layers are present, which train more abstract representations as you head up. Figure 5.3 depicts the training of neural network [35].

Deep Neural Network (DNN) has superior performance in image identification, virtual art processing, natural language processing, weather forecasting, drug discovery, military services, and so on [36].

5.2.2 Advantages of Deep Learning

The advantages of deep learning methods are as follows [37]:

■ Features are trained automatically and tuned optimally to achieve desired outcome.
■ DNN scan be adapted to diversified problems in the future.
■ Similar DNN framework can be best suited in distinguished domains and data types.

Deep learning, however, requires specialized computer hardware and is extremely expensive to train due to complex data models.

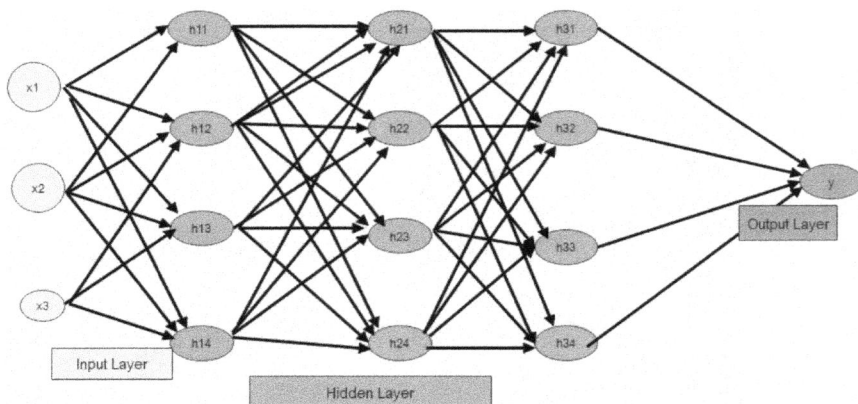

Figure 5.3   Architecture of Deep Neural Network (DNN) [29].

5.2.3 Applications of Deep Learning

DNN is a key enabler in handling numerous big data, so it's ubiquitous; for example, it is utilized in healthcare for diagnosing a disease considering symptoms, lab measurements, test results, DNA tests, etc. as inputs to achieve the desired output may be one of a set of possible diseases.

In the context of the smart grid, DNN methods enable efficient energy management for optimum utilization of the renewable energy resources to overcome the intermittent nature of utility resources and sustain demand response [4].

5.3 Features of Nanogrid

5.3.1 Integration of DGs in Interconnected Grid

Distributed generation (DG) plays a critical role in the distribution system. DG enhances system capability, reduces line losses, mitigates congestions, improves reliability, minimizes reserve limits requisite at minimum cost, and reduces power blackouts [15, 16, 25]. Falvo and Martirano and Brenna et al. investigated the concept and merits of interconnected microgrids. The fundamental theoretical approach is that indirectly, subsystems may act as an auxiliary power source for the interconnected grid [38]. Due to its distinguished characteristics, lots of economic merits can be accomplished by acceptable controlling of DG power cost, which could lead

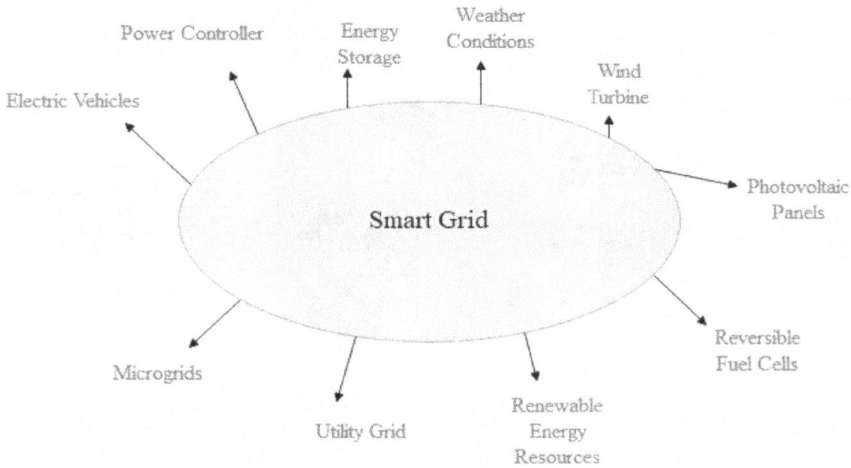

Figure 5.4   Integration of DGs in Interconnected Grid [43].

to massive cost reductions. Figure 5.4 depicts the integration of DGs with the inter-connected grid.

Regardless of the aforementioned preferences for the NG, grid interconnection has its impediments of complexity in interoperability, communications and information technology must be integrated and enormous protective devices must be coordinated to ensure the reliability of the overall grid [39–42].

5.3.2 Characteristics of Islanded Nano-grid

Islanded NGs definitely prove a better substitute for household demand in faraway or distant areas. Compact NGs do not have an essential bigger alternate to stand-ing apparatus, not like AC and DC nanogrids. The nanogrid centralized the power converter at one point. Figure 5.5 depicts the coordination between the generation, storage, and utilization of electric energy through Islanded Nanogrid. [44].

In isolated nanogrid, power converters will manipulate voltages to interface gen-erating sources, energy storage, and loads. The redundant structure enables Islanded Nanogrid to be more resilient to power abnormalities. In [46] the architecture of Keating Nanogrid installation at the Illinois Institute of Technology (IIT) Chicago demonstrated that the building was operated as a stand-alone power system, which was competent to resolve issues in case of any adversities.

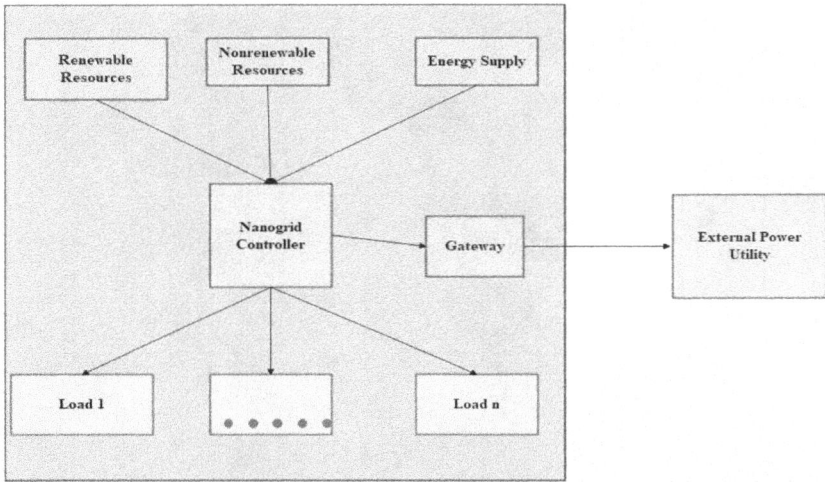

Figure 5.5 The Nanogrid for Offline Grid System [45].

5.4 Energy Management of Nanogrid with the Essence of Deep Learning

Energy management is the prime concern to ensure the efficacy and reliability of the nanogrid. In the articles [47, 48] investigation has been carried out to enhance energy management through optimization techniques. However, more emphasis was given on the energy management system to overcome the intermittent nature of renewable energy. Moreover, the energy management system (EMS) can be explained as an exhaustive self-acting and concurrent system applied for flawless schedule and management of DERs and governable loads, employed within an electrical division system.

With the implementation of EMS in nanogrid, data administration, grid statistics, regulation, and guidance over all the automated DGs and ESS will become seamless. A typical EMS requires huge data input such as power cost, state of charge (SoC) of the ESS, implementation schedule, certainty and safety constraint of the network, details at the point of common coupling (PCC) to install a set of regulated actions [49]. The EMS has to monitor the huge input and output data as shown in Figure 5.6. Despite that, the concept of big data analytics in nanogrid plays a very prominent role in the pervasive application.

Power quality issues in the nanogrid such as harmonics, sags, swells, transients, fluctuations, power outage, notching, power factor disturbances, etc. can be resolved by power line conditioners, for example, filters and automatic voltage regulators to enhance the system power quality. Despite that, they are not capable of resolving blackout. So, to overcome the intermittent nature of renewable resources, supply side management (SSM) can be achieved with a novel UPS strategy that can result as a power conditioner, which has been described by F. Iov et al. [51]. To tackle with

Figure 5.6   Input and Output Data for Energy Management System [50].

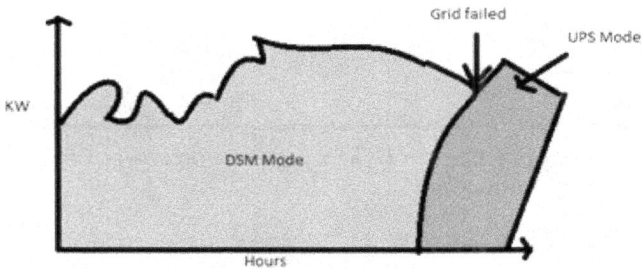

Figure 5.7   UPS/DSM Mode [53].

the load variation, DSM technique prefers crest shaving, gap filling, and load transpose to manage load management [52]. SSM/DSM operation modes are shown in Figure 5.7 [53].

In Figure 5.7, the system will operate as a DSM mode during normal operation and switch to UPS mode in case of deficits of grid by delivering energy to the load end and avoid any interference of power.

Integrated energy management system for PV storage hybrid system regulates by an energy management controller, which adopts a solar panel that delivers energy to a BES unit and a dc bus. The BES unit is connected to the DC bus through a DC-Link converter, which can reserve energy through the solar PV panels or discharge to fulfill the domestic demand. As a result, domestic load is forwarded through the PV Panel, the grid, and the BES unit as shown in Figure 5.8. They also formulated the intelligent approach of stochastic energy management in three distinguished steps for efficient energy management [54].

The description of each step is illustrated as follows:

■ Step 1: Load and PV generation forecast (duration 15 interval).
■ Step 2: Maximization on receding horizon.
■ Step 3: Rule-based control.

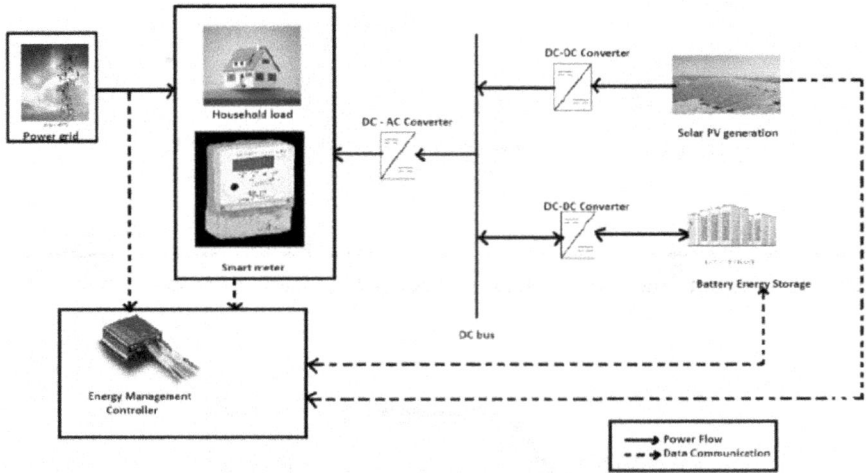

Figure 5.8 Domestic PV-Reserve Hybrid System with Power Converters and Energy Management Controller [54].

In step 1, the authors recommended the feature selection and forecasting algorithm to find optimum utilization of the grid, enhance veracity of the prediction of updated load and PV generation forecasting. They preferred dynamic learning approach of long short-term memory (LSTM) method, which is based on RNN for forecasting data.

In step 2, as the prediction of the updated load and PV generation does not exactly match the actual scenario, to achieve accuracy, uncertainties have been incorporated for decision-making capabilities. Therefore, a correlation structure between generation and load demand has been validated.

In step 3, authors revealed that only charging/discharging of the BES is not sufficient for accurate planning of the generation. So, a rule-based controller is also incorporated with the optimization method to update optimal set points of PV generation and thus avoid energy losses.

Additionally, the energy management of an NG is construct based on regulation schemes. To label the energy division problem in NGs, Lyapunov optimization to decide the amount of energy division for each NG in actual-time fashion was described in the manuscript [19, 20, 55, 56]. Control strategy allows the source to communicate with end-users through local information. As introduced in articles [57–60], adaptive droop control technique for load balancing, with the features of "plug & play" functionality, has its limitations of voltage drop in the grid and a poor current sharing performance.

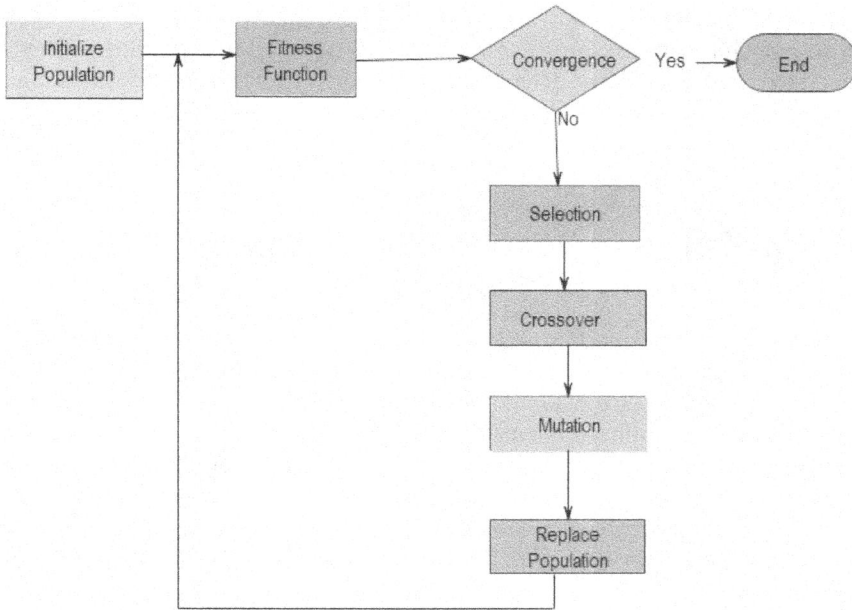

Figure 5.9   Flowchart of Genetic Algorithm (GA) [67].

Artificial intelligence is a global optimization search process. Artificial neural network (ANN) is a numerical-free estimator, which can approximate how the output function can be based on the input in the absence of the need for compound mathematical models [61–65]. Control structure of ANN is a procedure to adjust these values so that the ANN can map all the input control values to the correlate with output control values. Genetic algorithm (GA) is motivated by the mechanism of natural choice where stronger individuals would likely be the winners in a engaged environment. The flowchart of GA is shown in Figure 5.9 [66, 67].

For efficient and intuitive decisions to be declared by EMS for continuous-variable problems, classical GA is not appropriate as it is binary encoded. Therefore, the concept of deep learning has been adopted, which keeps away from the binary encoding and decoding of continuous variables to solve complicated problems of nanogrids that have not yet been solved by machine learning (ML) methods. Common applications of machine learning include Amazon Alexa, Google translate self-driving car, fraud detection, etc. [32, 68].

Deep learning (DL) models have excellent performance on huge data sets. They use neural network structure, that is why they are often referred to as deep neural networks. The term "deep" usually symbolizes the number of invisible layers in the neural network. The traditional neural network only has inherent two to three hidden layers, while deep networks can have as many as 150. Deep learning models are trained by using large sets of labeled data and neural network architectures that have been features straight from the data without the need for manual feature extraction [69–71].

5.5 Challenges and Solutions of Energy Management in Nanogrid

The current power management system (EMS) is based on traditional production technologies and submissive consumers. However, in the smart paradigm, the framework to provide EMS should change significantly to address the customer's awareness on power utilization and their involvement. Hence, the developed EMS through deep learning algorithm will enhance efficiency, reliability, stability, and accuracy.

There are various challenges associated with the integration of EMS in the nanogrid. One of the major issues is to handle a large volume of available data at EMS for electricity consumption. Load utilization and weather prediction are the critical information for the EMS. For a well-organized operation of the nanogrid, accurate forecasted data plays an important role. If load utilization or production values are underestimated, then the grid operation and stability will be at risk due to the insufficient spinning reserve of the online units. Conversely, if generation values are fabricated, many units will be dispatched; thereafter, the operational cost will increase. Here upon an efficient data processing algorithm and effective computing assets process such large volumes in an appropriate amount of time [72–75].

Moving forward, another challenge corresponds to the identification of the exact problem available from the data at the EMS. The factors include environmental, economic, and social that increase the complication of the EMS. The third challenge is the integration of renewable energy resources with the demand side management (DSM) at the EMS.

This section plays the interface between the grid and the end-users to define a suitable schedule for the governable loads. Feasible loads can be tabulated into two groups as curtail able load or shedding load. Curtail able loads are those that can be uncoupled at any time implying a certain cost and whose depletion can be specified by parameters such as the percentage of load curtailment, the maximum frequency of curtailments, among others. Whereas the shedding loads are those whose consumption can be rescheduled and that can be specified by parameters such as large length of rescheduling time, cost of rescheduling, among others. DSM adds flexibility to the operation of EMS if an efficient data processing algorithm is incorporated [76].

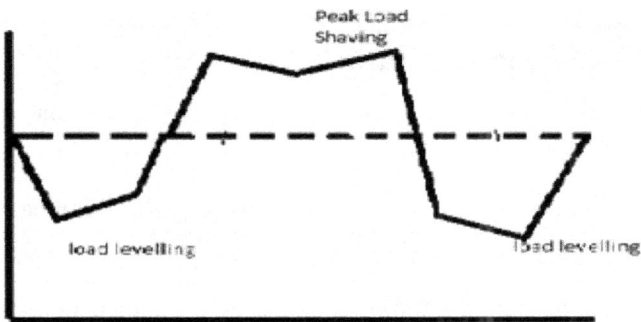

Figure 5.10 The Principle of Load Leveling [52].

The next challenge is to maintain load level and peak shift. They are widely used to mitigate load fluctuations and improve power quality. Load leveling is the important function of the EMS. It is useful to reduce the influence of the load variation and lower the cost of the nanogrid. The principle of peak shaving and load-leveling is shown in Figure 5.10.

In Figure 5.10, the real load is changing constantly due to switching on and off the devices in the system. With the help of the EMS, the sudden increase and decrease of load can be compensated for, to work ESS like a floating load. Therefore, the upcoming curve will behave smoothly for that duration, which results mean the value of the load does not change too much [58, 59]. Hence, the variation of the load that is compensated by EMS proves to be improved in power quality.

In the nanogrid, EMS either can function as a load (during the charging period) or as a generator (when discharging). Therefore, it can perform peak shifting function to reduce or eliminate the peaks and valleys in the load profile, so the nanogrid can satisfy the high demand during peak time. Thereafter, the EMS is provided with the information of how much energy or power is needed during peak times. Thereafter, the EMS can store the required energy during the valley period and support the system in the peak time.

Typically, peak shifting is based on load forecasting technology. This is a big challenge to ensure the accuracy of the forecasted data because of the contribution of uncertain factors in the smart grid paradigm. This is based on the assumption of the environmental factors and electricity demand [77–79].

Overall, an efficient processing and computing algorithm is required to achieve accuracy, improved power quality, ensure reliability, and maintain the stability of the smart grid paradigm [80].

5.6 Comparison between Deep Learning and Machine Learning

The contrast between deep learning and machine learning based on renowned features is as follows:

Table 5.1 Comparison between Machine Learning and Deep Learning

S. No	Features	Machine Learning (ML)	Deep Learning (DL)
1	Data Dominion [29].	Excellent presentation on a small/medium dataset.	Excellent presentation on huge dataset.
2	Feature Engineering [69].	Needs to recognize the features that constitute the data.	No need to recognize the best feature that constitute the data.
3	Hardware Dependencies [68, 70].	Work on low end machine.	Requires powerful machine.
4	Execution Time [79].	From few minutes to hours.	Upto weeks, neural network needs to compute a significant number of weights.
5	Interpretability [80].	Some algorithms are effortless to interpret (logistic, decision tree).	Hard to impossible.

Table 5.1 presents the leverage of deep learning over machine learning in the aspect that machine learning is not capable enough to manage huge dimensional data for continuously varying input and output.

5.7 Conclusion and Future Scope

DL has attracted a great deal of attention all over the world in recent years. In this chapter, we have studied the energy management system (EMS) of a nanogrid. The chapter focused on the challenges and solutions of implementing the EMS in the smart grid paradigm. It also revealed the control strategies for energy management systems and presented the importance of DL over ML for massive datasets.

Smart grids, microgrids, and nanogrids are the greatest potential application areas of DL. This algorithm has significant application in the future for smart grids. Only a few researchers have investigated this field in recent years. This is just the beginning, and many more milestones need to be covered to resolve problems in the depth of the smart grid paradigm. Still, it is an energizing factor for all. They can be applied in prediction, anomaly decision, decision-making support for control, etc. Furthermore, for specific application, deep learning algorithm is preferable. The research can be broadened in various ways.

References

1. E. Esenogho, K. Djouani, and A. Kurien, "Survey of trends challenges and prospect," *IEEE Access*, p. 1, 2022, doi: 10.1109/ACCESS.2022.3140595.
2. L. Atzori, A. Iera, and G. Morabito, "The internet of things: A survey," *Comput. Networks*, vol. 54, no. 15, pp. 2787–2805, 2010, doi: 10.1016/j.comnet.2010.05.010.
3. A. Zanella, N. Bui, A. Castellani, L. Vangelista, and M. Zorzi, "Internet of things for smart cities," *IEEE Internet Things J.*, vol. 1, no. 1, pp. 22–32, 2014, doi: 10.1109/JIOT.2014.2306328.
4. W. He, "Load forecasting via deep neural networks," *Procedia Comput. Sci.*, vol. 122, pp. 308–314, 2017, doi: 10.1016/j.procs.2017.11.374.
5. T. L. Nguyen, J. M. Guerrero, and G. Griepentrog, "A self-sustained and flexible control strategy for islanded DC nanogrids without communication links," *IEEE J. Emerg. Sel. Top. Power Electron.*, vol. 8, no. 1, pp. 877–892, 2020, doi: 10.1109/JESTPE.2019.2894564.
6. S. M. Cialdea, J. A. Orr, A. E. Emanuel, and T. Zhang, "An optimal battery energy storage charge/discharge method," *2013 IEEE Electr. Power Energy Conf. EPEC 2013*, 2013, doi: 10.1109/EPEC.2013.6802908.
7. R. Arun, K. S. Mohammed Gohar Latheef, and G. Anandhakumar, "Grid interconnection of renewableenergy sources at the distribution level with power-quality improvement features," *Int. J. Appl. Eng. Res.*, vol. 10, no. 33 Special Issue, pp. 25622–25626, 2015, doi: 10.23883/ijrter.conf.20171216.014.uylw8.
8. D. O. Akinyele and R. K. Rayudu, "Review of energy storage technologies for sustainable power networks," *Sustain. Energy Technol. Assessments*, vol. 8, pp. 74–91, 2014, doi: 10.1016/j.seta.2014.07.004.
9. R. P. S. Chandrasena, F. Shahnia, A. Ghosh, and S. Rajakaruna, "Operation and control of a hybrid AC-DC nanogrid for future community houses," *2014 Australas. Univ. Power Eng. Conf. AUPEC 2014—Proc.*, October, pp. 1–6, 2014, doi: 10.1109/AUPEC.2014.6966617.
10. A. Sannino, G. Postiglione, and M. H. J. Bollen, "Feasibility of a dc network for commercial facilities," *Conf. Rec. Annu. Meet. (IEEE Ind. Appl. Soc.*, vol. 3, pp. 1710–1717, 2002, doi: 10.1109/IAS.2002.1043764.
11. E. Planas, J. Andreu, J. I. Gárate, I. Martínez De Alegría, and E. Ibarra, "AC and DC technology in microgrids: A review," *Renew. Sustain. Energy Rev.*, vol. 43, pp. 726–749, 2015, doi: 10.1016/j.rser.2014.11.067.

12. N. L. Diaz, T. Dragicevic, J. C. Vasquez, and J. M. Guerrero, "Intelligent distributed generation and storage units for DC microgrids—A new concept on cooperative control without communications beyond droop control," *IEEE Trans. Smart Grid*, vol. 5, no. 5, pp. 2476–2485, 2014, doi: 10.1109/TSG.2014.2341740.

13. J. A. Sa'Ed, N. Ismail, S. Favuzza, M. G. Ippolito, and F. Massaro, "Effect of voltage deviations on power distribution losses in presence of DG technology," *2015 Int. Conf. Renew. Energy Res. Appl. ICRERA 2015*, vol. 5, pp. 766–771, 2015, doi: 10.1109/ICRERA.2015.7418515.

14. J. A. Sa'Ed, S. Favuzza, M. G. Ippolito, and F. Massaro, "Integration issues of distributed generators considering faults in electrical distribution networks," *Energycon 2014—IEEE Int. Energy Conf.*, pp. 1062–1068, 2014, doi: 10.1109/ENERGYCON.2014.6850556.

15. J. A. Sa'ed, M. Quraan, Q. Samara, S. Favuzza, and G. Zizzo, "Impact of integrating photovoltaic based DG on distribution network harmonics," *Conf. Proc.—2017 17th IEEE Int. Conf. Environ. Electr. Eng. 2017 1st IEEE Ind. Commer. Power Syst. Eur. EEEIC/I CPS Eur. 2017*, pp. 1–5, 2017, doi: 10.1109/EEEIC.2017.7977786.

16. J. A. Sa'Ed, S. Favuzza, M. G. Ippolito, and F. Massaro, "Investigating the effect of distributed generators on traditional protection in radial distribution systems," *2013 IEEE Grenoble Conf. PowerTech, POWERTECH 2013*, 2013, doi: 10.1109/PTC.2013.6652100.

17. J. A. Sa'ed, M. Amer, A. Bodair, A. Baransi, S. Favuzza, and G. Zizzo, "A simplified analytical approach for optimal planning of distributed generation in electrical distribution networks," *Appl. Sci.*, vol. 9, no. 24, 2019, doi: 10.3390/app9245446.

18. J. A. Sa'ed, M. Amer, A. Bodair, A. Baransi, S. Favuzza, and G. Zizzo, "Effect of integrating photovoltaic systems on electrical network losses considering load variation," *Proc.—2018 IEEE Int. Conf. Environ. Electr. Eng. 2018 IEEE Ind. Commer. Power Syst. Eur. EEEIC/I CPS Eur. 2018*, 2018, doi: 10.1109/EEEIC.2018.8494433.

19. S. Ganesan, . . . D. P.-I. transactions on, and undefined 2015, "Control scheme for a bidirectional converter in a self-sustaining low-voltage DC nanogrid," *ieeexplore.ieee.org*, Accessed: Jan. 05, 2021. [Online]. Available: https://ieeexplore.ieee.org/abstract/document/7088627/.

20. N. Liu, X. Yu, W. Fan, C. Hu, . . . T. R.-I. T. on, and undefined 2017, "Online energy sharing for nanogrid clusters: A Lyapunov optimization approach," *ieeexplore.ieee.org*, Accessed: Jan. 05, 2021. [Online]. Available: https://ieeexplore.ieee.org/abstract/document/7845698/.

21. A. Salazar, A. Berzoy, W. Song, and J. M. Velni, "Energy management of islanded nanogrids through nonlinear optimization using stochastic dynamic programming," *IEEE Trans. Ind. Appl.*, vol. 56, no. 3, pp. 2129–2137, 2020, doi: 10.1109/TIA.2020.2980731.

22. S. C. Joseph, A. Mohammed Ajlif, P. R. Dhanesh, and S. Ashok, "Smart power management for DC nanogrid based building," *2018 IEEE Recent Adv. Intell. Comput. Syst. RAICS 2018*, no. 1, pp. 142–146, 2019, doi: 10.1109/RAICS.2018.8635070.

23. E. Rodriguez-Diaz, J. C. Vasquez, and J. M. Guerrero, "Intelligent DC homes in future sustainable energy systems: When efficiency and intelligence work together," *IEEE Consum. Electron. Mag.*, vol. 5, no. 1, pp. 74–80, 2016, doi: 10.1109/MCE.2015.2484699.

24. H. Kakigano, Y. Miura, and T. Ise, "Low-voltage bipolar-type dc microgrid for super high quality distribution," *IEEE Trans. Power Electron.*, vol. 25, no. 12, pp. 3066–3075, 2010, doi: 10.1109/TPEL.2010.2077682.

25. A. C. Luna, N. L. Diaz, L. Meng, M. Graells, J. C. Vasquez, and J. M. Guerrero, "Generation-side power scheduling in a grid-connected DC microgrid," *2015 IEEE 1st Int. Conf. Direct Curr. Microgrids, ICDCM 2015*, pp. 327–332, 2015, doi: 10.1109/ICDCM.2015.7152063.

26. W. Dalbon, S. Leva, M. Roscia, and D. Zaninelli, "Hybrid photovoltaic system control for enhancing sustainable energy," *Proc. IEEE Power Eng. Soc. Transm. Distrib. Conf.*, vol. 1, Summer, pp. 134–139, 2002, doi: 10.1109/pess.2002.1043198.

27. J. Schönberger, R. Duke, and S. D. Round, "DC-bus signaling: A distributed control strategy for a hybrid renewable nanogrid," *IEEE Trans. Ind. Electron.*, vol. 53, no. 5, pp. 1453–1460, 2006, doi: 10.1109/TIE.2006.882012.

28. M. Zekić-Sušac, S. Mitrović, and A. Has, "Machine learning based system for managing energy efficiency of public sector as an approach towards smart cities," *Int. J. Inf. Manage.*, vol. 58, 2021, doi: 10.1016/j.ijinfomgt.2020.102074.

29. A. S. Lundervold and A. Lundervold, "An overview of deep learning in medical imaging focusing on MRI," *Z. Med. Phys.*, vol. 29, no. 2, pp. 102–127, 2019, doi: 10.1016/j.zemedi.2018.11.002.

30. A. I. Dounis, "Artificial intelligence for energy conservation in buildings," *Adv. Build. Energy Res.*, vol. 4, no. 1, pp. 267–299, 2010, doi: 10.3763/aber.2009.0408.

31. W. He, "Load Forecasting via Deep Neural Networks," *Procedia Computer Science*, vol. 122, pp. 308–314, 2017, doi: 10.1016/j.procs.2017.11.374.

32. Q. Song, Y. Wu, and Z. Liu, "An overview of deep learning in power production," *PervasiveHealth Pervasive Comput. Technol. Healthc.*, September, 2019, doi: 10.1145/3386415.3387080.

33. N. Bassiliades and G. Chalkiadakis, "Artificial intelligence techniques for the smart grid," *Adv. Build. Energy Res.*, vol. 12, no. 1, pp. 1–2, 2018, doi: 10.1080/17512549.2017.1410362.

34. M. Cullell-Dalmau, M. Otero-Viñas, and C. Manzo, "Research techniques made simple: Deep learning for the classification of dermatological images," *J. Invest. Dermatol.*, vol. 140, no. 3, pp. 507–514.e1, 2020, doi: 10.1016/j.jid.2019.12.029.

35. P. S. Sarkar, "Deep neural Network." p. Lecture 33, 2016, [Online]. Available: https://nptel.ac.in/content/storage/106/105/106105152/MP4/mod01lec01.mp4.

36. A. Stuhlsatz, J. Lippel, and T. Zielke, "Discriminative feature extraction with deep neural networks," *Proc. Int. Jt. Conf. Neural Networks*, no. 4, 2010, doi: 10.1109/IJCNN.2010.5596329.

37. "Deep Learning," *Wikipedia Contributors.* [Online]. Available: https://en.wikipedia.org/w/index.php?title=Deep_learning&oldid=1063595570.

38. D. C. Microgrids, "Conceptual study for open energy system: Distributed energy network using interconnected," *IEEE Trans. Smart Grid*, vol. 6, no. 4, July 2015, 1999.

39. P. Systems, "1547 IEEE standards," *IEEE Std 1547–2003*, no. July, 2003.

40. M. F. Shaaban, Y. M. Atwa, and E. F. El-Saadany, "DG allocation for benefit maximization in distribution networks," *IEEE Trans. Power Syst.*, vol. 28, no. 2, pp. 639–649, 2013, doi: 10.1109/TPWRS.2012.2213309.

41. Y. A. Kumar, "Units in radial distribution systems using nature," *2018 Int. Conf. Power, Energy, Control Transm. Syst.*, no. 3, pp. 1–6, 2018.

42. M. C. V. Suresh and E. J. Belwin, "Optimal DG placement for benefit maximization in distribution networks by using Dragonfly algorithm," *Renewables Wind. Water, Sol.*, vol. 5, no. 1, 2018, doi: 10.1186/s40807-018-0050-7.

43. M. Bajaj and A. K. Singh, "Grid integrated renewable DG systems: A review of power quality challenges and state-of-the-art mitigation techniques," *Int. J. Energy Res.*, vol. 44, no. 1, pp. 26–69, 2020, doi: 10.1002/er.4847.

44. S. C. Joseph, S. Ashok, and P. R. Dhanesh, "An effective method of power management in DC nanogrid for building application," *2017 IEEE Int. Conf. Signal Process. Informatics, Commun. Energy Syst. SPICES 2017*, pp. 1–5, 2017, doi: 10.1109/SPICES.2017.8091303.

45. D. Burmester, R. Rayudu, W. Seah, and D. Akinyele, "A review of nanogrid topologies and technologies," *Renewable and Sustainable Energy Reviews*, vol. 67, 2017, doi: 10.1016/j.rser.2016.09.073.

46. M. Shahidehpour, Z. Li, . . . W. G.-I. E., and undefined 2017, "A Hybrid ac\/dc Nanogrid: The Keating Hall Installation at the Illinois Institute of Technology," *ieeexplore.ieee.org*, Accessed: Jan. 05, 2021. [Online]. Available: https://ieeexplore.ieee.org/abstract/document/7942242/.

47. M. Rafiee Sandgani and S. Sirouspour, "Energy management in a network of grid-connected microgrids/nanogrids using compromise programming," *IEEE Trans. Smart Grid*, vol. 9, no. 3, pp. 2180–2191, 2018, doi: 10.1109/TSG.2016.2608281.

48. P. Arboleya *et al.*, "Efficient energy management in smart micro-grids: ZERO grid impact buildings," *IEEE Trans. Smart Grid*, vol. 6, no. 2, pp. 1055–1063, 2015, doi: 10.1109/TSG.2015.2392071.

49. L. Mariam, M. Basu, and M. F. Conlon, "Microgrid: Architecture, policy and future trends," *Renew. Sustain. Energy Rev.*, vol. 64, pp. 477–489, 2016, doi: 10.1016/j.rser.2016.06.037.

50. M. Meliani, A. El Barkany, I. El Abbassi, A. M. Darcherif, and M. Mahmoudi, "Energy management in the smart grid: State-of-the-art and future trends," *Int. J. Eng. Bus. Manag.*, vol. 13, pp. 1–26, 2021, doi: 10.1177/18479790211032920.

51. F. Iov, M. Ciobotaru, D. Sera, R. Teodorescu, and F. Blaabjerg, "Power electronics and control of renewable energy systems," *Proc. Int. Conf. Power Electron. Drive Syst.*, December, 2007, doi: 10.1109/PEDS.2007.4487668.

52. C. W. Gellings, "The concept of demand-side management for electric utilities," *Proc. IEEE*, vol. 73, no. 10, pp. 1468–1470, 1985, doi: 10.1109/PROC.1985.13318.

53. M. Ashari, W. W. L. Keerthipala, and C. V. Nayar, "A single phase parallely connected uninterruptible power supply/demand side management system," *IEEE Trans. Energy Convers.*, vol. 15, no. 1, pp. 97–102, 2000, doi: 10.1109/60.849123.

54. F. Hafiz, M. A. Awal, A. R. De Queiroz, and I. Husain, "Real-time stochastic optimization of energy storage management using deep learning-based forecasts for residential PV applications," *IEEE Trans. Ind. Appl.*, vol. 56, no. 3, pp. 2216–2226, 2020, doi: 10.1109/TIA.2020.2968534.

55. C. Sun, F. Sun, and S. J. Moura, "Data enabled predictive energy management of a PV-battery smart home nanogrid." Accessed: Jan. 05, 2021. [Online]. Available: https://ieeexplore.ieee.org/abstract/document/7170867/.

56. W. Fan, N. Liu, and J. Zhang, "An event-triggered online energy management algorithm of smart home: Lyapunov optimization approach," *Energies*, vol. 9, no. 5, 2016, doi: 10.3390/en9050381.

57. S. I. Gkavanoudis, K. O. Oureilidis, and C. S. Demoulias, "An adaptive droop control method for balancing the SoC of distributed batteries in AC microgrids," *2016 IEEE 17th Work. Control Model. Power Electron. COMPEL 2016*, vol. 2016, 2016, doi: 10.1109/COMPEL.2016.7556698.

58. X. Hou, Y. Sun, W. Yuan, H. Han, C. Zhong, and J. M. Guerrero, "Conventional P-ω/Q-V droop control in highly resistive line of low-voltage converter-based AC microgrid," *Energies*, vol. 9, no. 11, 2016, doi: 10.3390/en9110943.
59. R. Majumder, A. Ghosh, G. Ledwich, and F. Zare, "Angle droop versus frequency droop in a voltage source converter based autonomous microgrid," *2009 IEEE Power Energy Soc. Gen. Meet. PES '09*, 2009, doi: 10.1109/PES.2009.5275987.
60. R. Majumder, G. Ledwich, A. Ghosh, S. Chakrabarti, and F. Zare, "Droop control of converter-interfaced microsources in rural distributed generation," *IEEE Trans. Power Deliv.*, vol. 25, no. 4, pp. 2768–2778, 2010, doi: 10.1109/TPWRD.2010.2042974.
61. E. Mocanu, P. H. Nguyen, M. Gibescu, and W. L. Kling, "Deep learning for estimating building energy consumption," *Sustain. Energy, Grids Networks*, vol. 6, no. xxxx, pp. 91–99, 2016, doi: 10.1016/j.segan.2016.02.005.
62. X. J. Luo and K. F. Fong, "Development of integrated demand and supply side management strategy of multi-energy system for residential building application," *Appl. Energy*, vol. 242, pp. 570–587, May 2019, doi: 10.1016/J.APENERGY.2019.03.149.
63. C. Phurailatpam, B. S. Rajpurohit, and L. Wang, "Planning and optimization of autonomous DC microgrids for rural and urban applications in India," *Renew. Sustain. Energy Rev.*, vol. 82, September 2017, pp. 194–204, 2018, doi: 10.1016/j.rser.2017.09.022.
64. X. Wu, X. Hu, Y. Teng, S. Qian, and R. Cheng, "Optimal integration of a hybrid solar-battery power source into smart home nanogrid with plug-in electric vehicle," *J. Power Sources*, vol. 363, pp. 277–283, 2017, doi: 10.1016/j.jpowsour.2017.07.086.
65. R. Kaur, V. Krishnasamy, and N. K. Kandasamy, "Optimal sizing of wind–PV-based DC microgrid for telecom power supply in remote areas," *IET Renew. Power Gener.*, vol. 12, no. 7, pp. 859–866, 2018, doi: 10.1049/iet-rpg.2017.0480.
66. M. S. Ismail, M. Moghavvemi, and T. M. I. Mahlia, "Characterization of PV panel and global optimization of its model parameters using genetic algorithm," *Energy Convers. Manag.*, vol. 73, pp. 10–25, 2013, doi: 10.1016/j.enconman.2013.03.033.
67. I. G. Damousis, A. G. Bakirtzis, and P. S. Dokopoulos, "Network-constrained economic dispatch using real-coded genetic algorithm," *IEEE Trans. Power Syst.*, vol. 18, no. 1, pp. 198–205, 2003, doi: 10.1109/TPWRS.2002.807115.
68. D. Zhang, X. Han, and C. Deng, "Review on the research and practice of deep learning and reinforcement learning in smart grids," *CSEE J. Power Energy Syst.*, vol. 4, no. 3, pp. 362–370, 2018, doi: 10.17775/cseejpes.2018.00520.
69. J. Jiao, "Application and prospect of artificial intelligence in smart grid," *IOP Conf. Ser. Earth Environ. Sci.*, vol. 510, no. 2, 2020, doi: 10.1088/1755-1315/510/2/022012.
70. Y. Bengio, "Learning deep architectures for AI," *Found. Trends Mach. Learn.*, vol. 2, no. 1, pp. 1–27, 2009, doi: 10.1561/2200000006.
71. H. Jia, N. Djilali, X. Yu, H. D. Chiang, and G. Xie, "Computational science in smart grids and energy systems," *J. Appl. Math.*, vol. 2015, no. 1, pp. 2–4, 2015, doi: 10.1155/2015/326481.
72. M. E. Khodayar and H. Wu, "Demand forecasting in the smart grid paradigm: Features and challenges," *Electr. J.*, vol. 28, no. 6, pp. 51–62, 2015, doi: 10.1016/j.tej.2015.06.001.
73. R. Bo and F. Li, "Probabilistic LMP forecasting considering load uncertainty," *IEEE Trans. Power Syst.*, vol. 24, no. 3, pp. 1279–1289, 2009, doi: 10.1109/TPWRS.2009.2023268.

74. FERC, "2010 Assessment of demand response and advanced metering—staff report," *Fed. Energy Regul. Comm.*, vol. 74, no. 0022–3042 SB-IM, pp. 2–5, 2011, [Online]. Available: www.ferc.gov/legal/staff-reports/2010-dr-report.pdf%5Cnwww.ferc.gov/legal/staff-reports/09-07-demand-response.pdf.

75. T. Lv, Q. Yang, X. Deng, J. Xu, and J. Gao, "Generation expansion planning considering the output and flexibility requirement of renewable energy: The case of Jiangsu province," *Front. Energy Res.*, vol. 8, no. March, pp. 1–11, 2020, doi: 10.3389/fenrg.2020.00039.

76. A. Ghoraishi, "Recovering waste heat from diesel generator exhaust; an opportunity for combined heat and power generation in remote Canadian mines," *J. Clean. Prod.*, vol. 225, 10 July, pp. 785–805, 2019.

77. H. Zou and Y. Yang, "Combining time series models for forecasting," *Int. J. Forecast.*, vol. 20, no. 1, pp. 69–84, 2004, doi: 10.1016/S0169-2070(03)00004-9.

78. J. M. Bates, "The combination of forecasts," *Essays Econom.*, vol. 20, no. 4, pp. 391–410, 2010, doi: 10.1017/cbo9780511753961.021.

79. K. Methaprayoon, W. J. Lee, S. Rasmiddatta, J. Liao, and R. Ross, "Multi-stage artificial neural network short-term load forecasting engine with front-end weather forecast," *Conf. Rec.—Ind. Commer. Power Syst. Tech. Conf.*, vol. 43, no. 6, pp. 1410–1416, 2006, doi: 10.1109/icps.2006.1677297.

80. Z. Yang, F. Zhu, and F. Lin, "Deep-reinforcement-learning-based energy management strategy for supercapacitor energy storage systems in urban rail transit," *IEEE Trans. Intell. Transp. Syst.*, vol. 22, no. 2, pp. 1150–1160, 2021, doi: 10.1109/TITS.2019.2963785.

Chapter 6

An Elitism-Based SAMP-JAYA Algorithm for Optimal VA Loading of Unified Power Quality Conditioner

Swati Gade and Rahul Agrawal

Contents

DOI: 10.1201/9781003301820-6

6.1 Introduction

Nowadays in power systems, use of non-linear loads and incorporation of renewable energy resources with the grid using power electronic converters (PECs) cause power quality (PQ) issues such as unbalanced voltages, voltage sag/swell, compensation of reactive power, and harmonic currents [1, 2]. Interruption of critical loads due to these PQ problems causes economical losses and a reduction in the system power factor (PF). Maintaining the quality of the power supply in accordance with the specified standards is essential for the optimum efficiency of the power system [3, 4]. A custom power device named Unified Power Quality Conditioner (UPQC) was introduced in 1998 by Fujita and Akagi [5]. It is used in the distribution system for the compensation of all types of PQ issues. It is found best mitigating solution for critical-sensitive loads due to its features of simultaneous compensation of both voltage and current-related PQ issues [6]. It consists of back-to-back connected two PECs connected via a communal DC-Link as shown in Figure 6.1. The main

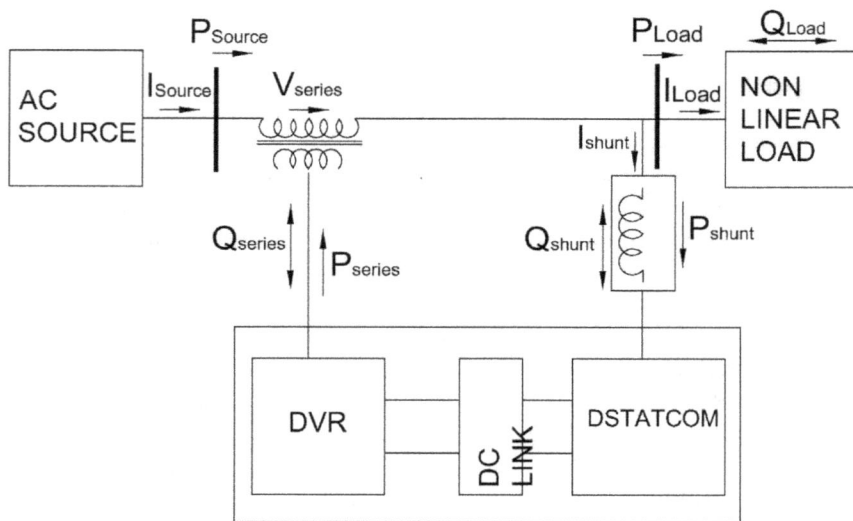

Figure 6.1 Block Diagram of UPQC.

function of DVR, that is, series-connected PEC is to compensate for voltage-related PQ issues. On the other hand, the PEC that is connected in parallel with the load is responsible to supply load reactive requirements and compensate for current-related PQ issues. It is called distributed STATCOM (DSTATCOM) [7].

The performance of UPQC is mainly dependent upon the control algorithm used for controlling it. The recent literature shows that researchers have tried to utilize UPQC effectively and efficiently by using various compensation methods [7–10]. Figure 6.2 shows the classification of UPQC based on its compensation method used.

The literature makes it clear that UPQC-S and UPQC-VAmin have higher utilization rates for both converters, which lowers the system's overall cost and makes the system more acceptable from an economical perspective [11].

From the literature, it is concluded that in the case of UPQC-VA$_{min}$ online/offline optimization approaches can be used to calculate an appropriate angle for injection of series voltage by DVR. In the offline optimization approach, this angle is calculated and a 2D lookup table is constructed. The optimal angle for DVR to inject series voltage is determined from this lookup table, and DVR is controlled accordingly [12, 13]. However, individual VA loading of converters is mainly dependent upon the operating conditions that are ignored in the literature studied. Various system factors, such as load PF, current, voltage sag/swell, source voltage THD, and load current THD, were taken into account when framing the optimization problem [14, 15]. Using the PSO technique, the optimal power angle value is calculated. To keep this ideal angle, the series voltage must be injected to keep the UPQC's VA loading to a minimum. However, the impact on individual PEC VA loadings as well as the series transformer is not taken into account. A phase angle control (PAC) approach-based technique was proposed by B. B. Ambati et al. [16] to maximize the overall UPQC's utilization. The correlation between the loadings of DVR and DSTATCOM and the angle δ is derived for different operating conditions. From this relationship, optimum angle δ is determined for minimum VA rating of UPQC for worst operating condition like voltage sag and swell. The variable PAC-based optimum design technique is proposed in Sa'ed et al. [17]. The variable PAC method was proposed in Sa'ed et al. [18]. To find the optimal VA ratings for DVR and DSTATCOM of UPQC that optimize the utilization of the entire UPQC system, a two-level control technique based on variable PAC is used. Series transformer rating is dependent upon the voltage injected for compensation. On the other hand, VA rating of the transformer and both converters will have an effect on each other [19]. Liu et al. [20] and Salazar et al. [21] attempted to manage the UPQC with the least amount of VA loading possible to decrease overall system losses under voltage sag circumstances. It has been noted that the optimization of the UPQC's VA loading has received little attention thus far. In Joseph et al. [22], PSO is used to find the optimum utilization of UPQC during voltage sag and steady-state conditions. The variable PAC approach is used to implement the TLBO [23] and JAYA [24] algorithms in order to determine the optimum VA loading for UPQC. Moreover

Compensation Methods of UPQC

UPQC-P
- Series voltage is injected by DVR in phase with source current
- Only active power for compensation

UPQC-Q
- Series voltage is injected by DVR in quadrature with source current
- Only reactive power for compensation
- Injected voltage leads the supply current therefore DVR required minimum VAr loading compared to UPQC-P
- It cannot be employed for voltage swell or voltage buck compensation

UPQC-S
- Series voltage is injected by DVR at some angle with source current
- Both active and reactive power required for compensation
- Some amount of load reactive power is supplied by DVR, reduces the burden on DSTATCOM

UPQC-VA$_{min}$
- Series voltage is injected by DVR at some optimum angle with source current
- Both active and reactive power required for compensation
- Some amount of load reactive power is supplied by DVR, reduces the burden on DSTATCOM

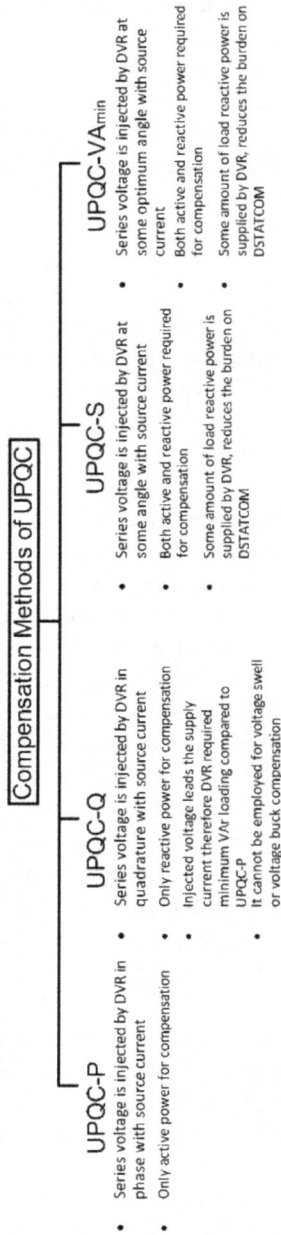

Figure 6.2   Classification of UPQC Based on Method of Compensation.

Luna et al., [25] presents the analysis of this method under steady-state and voltage sag conditions.

From the literature on engineering optimization, it is concluded that the solution to the complex engineering optimization problems in the shortest period is the primary challenge. Traditional approaches are not suitable for such complex problems, since they are monotonous, slow, and less accurate. These limitations encourage the researchers to develop metaheuristic-based computational techniques. These techniques require less information about the problem for achieving the global optimum solution [26]. While these optimization strategies have several benefits, they need fine-tuning of algorithm-specific parameters that are unique for each algorithm [27]. This is overcome by implementing algorithm-specific parameter-less techniques like TLBO [28] and the JAYA algorithm [29].

Advanced optimization approaches based on multi-population are utilized to boost the search diversity. In these techniques, the entire population is split into groups called sub-populations. The main advantage of these strategies is that problem changes can be identified effectively as sub-populations are distributed around the search space [30]. The merging and dividing technique is employed for interaction between the sub-populations when a change in the solution is recognized. For various problems, these techniques are more efficient and perform well than fixed population size techniques [31]. The size of sub-populations varies during searching. All these problems are addressed in the modified variants of the JAYA algorithm named self-adaptive multi-population JAYA (SAMP-JAYA) [30] and a self-adaptive multi-population elitist JAYA (SAMPE-JAYA) algorithm, which is an extension of SAMP-JAYA [32].

This chapter aims to discover the best possible power angle δ for various operating conditions for which VA loading of both converters is optimal and thus utilization rate of overall UPQC is maximized. For this, the SAMPE-JAYA optimization technique is used. The main objectives of this chapter are as follows:

1. To identify the appropriate power angle for every operational condition using the SAMPE-JAYA algorithm.
2. To investigate the performance of both the converters of UPQC for all operating conditions without negotiating any of its compensation proficiencies.
3. To examine the effectiveness of several optimization techniques for the optimal UPQC design based on VA loadings, including PSO, TLBO, and various JAYA algorithm versions.

The following is how the chapter is structured: The Problem Formulation was stated in section 2. Section 3 discusses the JAYA algorithm and its modifications in detail. The application of the SAMPE-JAYA algorithm for the investigation of optimum UPQC utilization is described in Section 4. Section 5 elaborates on the findings and discussion. Section 6 contains the final remarks and future assessments.

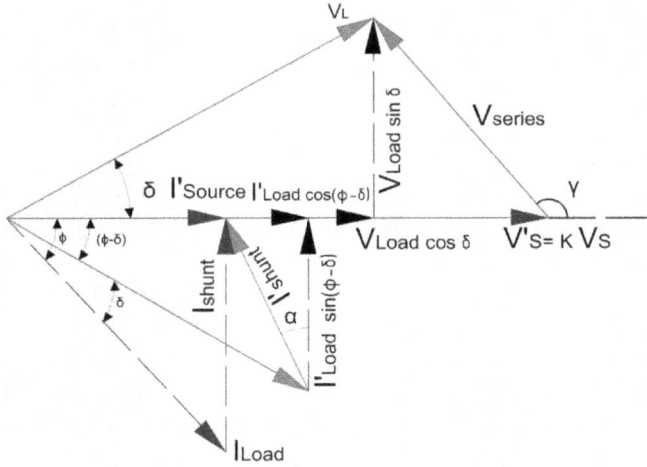

Figure 6.3   Phasor Diagram of UPQC Based on PAC during Voltage Sag.

6.2 Problem Formulation

Working of UPQC-VA$_{min}$ from the phasor diagram is explained in Figure 6.3. DVR injects series voltage V_{DVR} at an angle γ for maintaining the proper power angle δ between V_{Load}' and V_{source}'. To find the optimal loading of UPQC, its mathematical model is taken from Sa'Ed et al. [16].

6.2.1 Mathematical Modeling

The magnitude of the injected DVR voltage is given by Equation 6.1:

$$V_{DVR}(\delta, K) = \sqrt{(V_L \times \cos \delta - K \times V_s)^2 + (V_L \times \sin \delta)^2} \tag{6.1}$$

Total power handled by the DVR is represented next

$$S_{DVR}(\delta, K) = \sqrt{(P_{DVR}(\delta, K))^2 + (Q_{DVR}(\delta, K))^2} \tag{6.2}$$

Total power required by the series transformer

$$S_{Trans} = \max \text{ of} \left[\sqrt{(V_L \times \cos \delta - K \times V_s)^2 + (V_L \times \sin \delta)^2} \right]_{K=K_{min}}^{K=K_{max}} \times \frac{I_s}{K_{min}} \tag{6.3}$$

Total power handled by the DSTATCOM,

$$S_{DSTATCOM}(\delta, K) = \sqrt{(P_{DSTATCOM}(\delta, K))^2 + (Q_{DSTATCOM}(\delta, K))^2} \qquad (6.4)$$

Total loading of the UPQC,

$$S_{TotalUPQC}(\delta, K) = S_{DVR}(\delta, K) + S_{DSTATCOM}(\delta, K) \qquad (6.5)$$

6.2.2 Objective Function and Constraints

The objective function of this engineering optimization problem is formulated as follows. This research aims to determine the best way to use UPQC during the voltage sag and steady-state conditions so that efficacy of the system will be maximum.

$$\text{Min. } S_{TotalUPQC}(\delta, K) = S_{DVR}(\delta, K) + S_{DSTATCOM}(\delta, K) \qquad (6.6)$$

Subject to constraints,

$$\left. \begin{array}{l} S_{DVR}(\delta, K) \leq DVR \text{ rating} \\ S_{DSTATCOM}(\delta, K) \leq DSTATCOM \text{ rating} \\ V_{DVR}(\delta, K) \leq \text{Max voltage injected by series transformer} \end{array} \right\} \qquad (6.7)$$

where maximum series injected voltage is calculated from Equation 6.1 by putting $k=k_{max}$ and $\delta=45°$,

With design variable limit constraints,

$$\left. \begin{array}{l} 0 \leq \delta \leq 45° \\ 0.6 \leq K \leq 1.1 \end{array} \right\} \qquad (6.8)$$

6.3 JAYA Algorithm and Its Variants

R. Venkata Rao invented the "JAYA" algorithm, a swarm intelligence-based heuristic method for addressing limited and unconstrained optimization problems, in the year 2016 [29]. In the JAYA algorithm for each iteration, a solution moves nearer to the best solution and moves away from the worst solution. As a result, the search process is effectively intensified and diversified [33]. From the literature, it is concluded that to solve the various engineering optimization problem, different improved varieties of the JAYA optimization algorithm, such as the SA-JAYA algorithm [34], SAMP-JAYA algorithm [30], and SAMPE-JAYA algorithm [31], are developed and implemented.

6.3.1 Basic JAYA Algorithm

The basic JAYA algorithm generates initial populations X at random utilizing the upper and lower bounds of the process variables. Following that, each solution's design variable is stochastically modified as follows:

$$x_{jk}^{i+1} = x_{jk}^{i} + r_{1j}^{i}\left\{x_{jbest}^{i} - \left|x_{jk}^{i}\right|\right\} - r_{2j}^{i}\left\{x_{jworst}^{i} - \left|x_{jk}^{i}\right|\right\} \tag{6.9}$$

where, x_{jbest}^{i} - best solutions among the existing populations

x_{jworst}^{i} -worst solutions among the existing populations

i- Iteration
j- Variable
k- candidate solution
r_{1j}^{i}, r_{2j}^{i} - random generated in the range of [0, 1].

6.3.2 SA-JAYA Algorithm

SA-JAYA aims to solve the issue of selecting the appropriate population size for the proper application. In this, the population size is automatically determined by the algorithm. If there are D design variables, then the initial population size is calculated as $N=10*D$. For the next iteration, the population size is estimated:

$$N_{new} = round(N_{old} - r*(N_{old})) \tag{6.10}$$

where r-random generated in the range of [-0.5, 0.5].

The population size is controlled by using an elitism principle. Worst solutions are replaced with the best if the population size of the next iteration is higher than the current iteration. In contrast, only the best solutions are moved to the next iteration if the population size is less than the current iteration.

6.3.3 SAMP-JAYA Algorithm

The SAMP-JAYA algorithm divides the overall population into sub-populations that communicate with one another to modify the solution.

6.3.4 SAMPE-JAYA Algorithm

This is the improved form of the SAMP-JAYA algorithm. In this, the elite solutions are substituted for the worst solutions for populations with poor fitness values. The modifications to the basic JAYA are mentioned in Table 6.1.

Table 6.1   Modifications in SAMPE-JAYA Optimization Algorithm

Sr. No	Modification	Outcome
1	Division of the entire population into several groups or sub-populations.	Rather than converging in a specific region, the number of sub-populations distributes the solution across the search space.
2	Elite solutions replace the worst solutions.	The algorithm is supposed to produce the best solution and track changes in the problem landscape.
3	The number of sub-populations is adaptively modified.	Changing the number of sub-populations aids in locating the best solution and boosting the search process's exploration and diversification.
4	Duplicate solutions are replaced by newly generated solutions.	This maintains diversity and enhances the exploration procedure.

6.4   Implementation of SAMPE-JAYA for Optimum VA Loading of UPQC

For this research work, a balanced power system with the balanced harmonic free load of 10kW+j10kVAr is considered, which is connected to three-phase 400-V, 50-Hz supply. The UPQC is connected to the system to serve the following two functions:

1. For keeping the source side PF unity.
2. 40% compensation capability for voltage sag/swell.

Flow chart of the SAMPE-JAYA algorithm implemented to determine optimum VA loading of the UPQC is shown in Figure 6.4.

6.5   Results and Discussion

The proposed optimization algorithm has been written in MATLAB®. The simulation is run on a system that is balanced and free of harmonics. Table 6.2 and Table 6.3 list the system's parameters as well as the algorithms used.

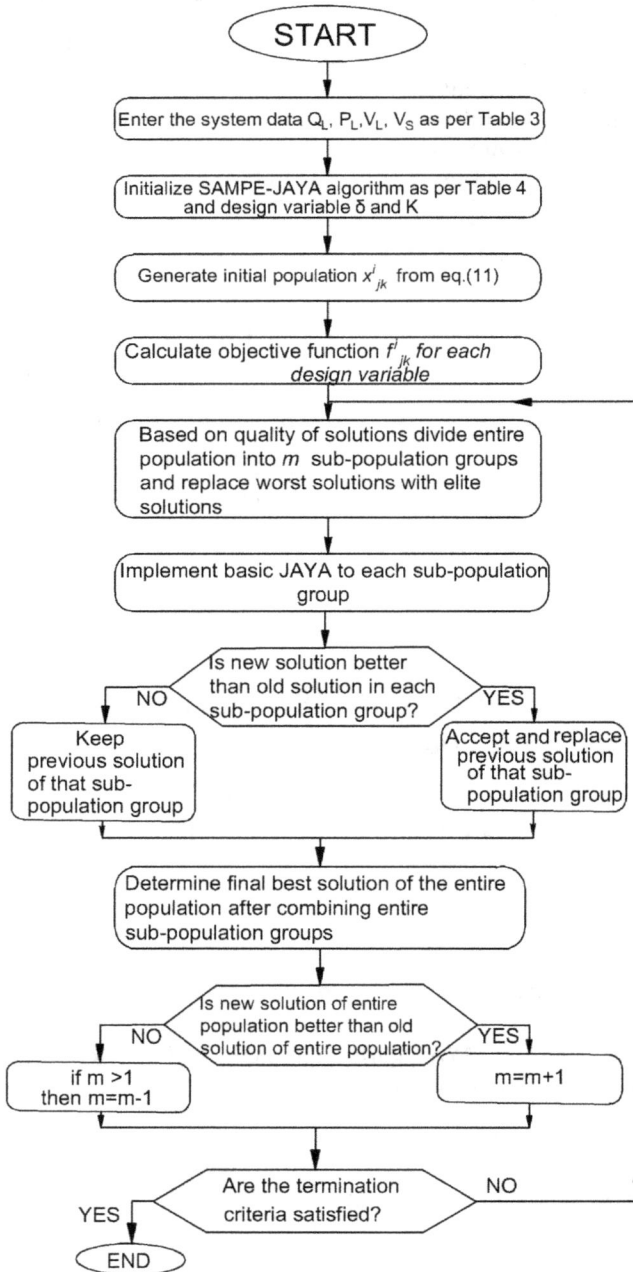

START

Enter the system data Q_L, P_L, V_L, V_S as per Table 3

Initialize SAMPE-JAYA algorithm as per Table 4 and design variable δ and K

Generate initial population x^i_{jk} from eq.(11)

Calculate objective function f^i_{jk} for each design variable

Based on quality of solutions divide entire population into m sub-population groups and replace worst solutions with elite solutions

Implement basic JAYA to each sub-population group

Is new solution better than old solution in each sub-population group?

NO

YES

Keep previous solution of that sub-population group

Accept and replace previous solution of that sub-population group

Determine final best solution of the entire population after combining entire sub-population groups

Is new solution of entire population better than old solution of entire population?

NO

YES

if m >1 then m=m-1

m=m+1

Are the termination criteria satisfied?

NO

YES

END

Figure 6.4 Flow Chart of the Implementation of SAMPE-JAYA Algorithm.

Table 6.2 System Parameters

Parameter	Quantity
Load	10kW+10kVAr
Source end voltage	400 Volt
Load end voltage	400 Volt
DVR rating	2490 VA
DSTATCOM rating	2907 VA
pu voltage during sag	0.6–0.9 pu
pu voltage during steady-state	0.9–1.1 pu

Table 6.3 Algorithm Parameters

	PSO	TLBO	Basic JAYA	SA-JAYA	SAMP-JAYA	SAMPE-JAYA
No. of Design Variables	2	--	2	2	2	2
Population Size	50	--	50	--	50	50
Max. No. of Iterations	50	50	50	50	50	50
Random	[0,1]	[0,1]	[0,1]	[-0.5,0.5]	[0,1]	[0,1]
Inertia Weight (w)	0.4 to 0.9	--	--	--	--	--
Acceleration Coefficient	2	--	--	--	--	--
No. of Subjects	--	2	--	--	--	--
Class Size	--	50	--	--	--	--
No. of Sub-population Groups	--	--	--	--	2	2
Elite Size for Case-1	--	--	--	--	--	12
Elite Size for Case-2	--	--	--	--	--	14

6.5.1 Analysis of VA Loading of UPQC without Optimization

Table 6.4 shows the UPQC VA loading without optimization. From the results, it is observed that for the UPQC-P approach, VA loading and voltage added by a series transformer for mitigation of 40% voltage sag is 2223 VA at 93 V. It is observed that the burden on DSTATCOM is more as compared to DVR in the case of UPQC-P, irrespective of operating condition, which results in increased loading of total UPQC. Compared with other operating conditions, it is found that during sag, the burden on overall UPQC is extreme. Similarly, with the UPQC-VA$_{min}$ approach, during sag, DVR shares the most power (3577 VA) at an angle of $\delta = 38°$, while DSTATCOM contributes the least (1048 VA). In both the approaches, entire load reactive power is supplied by DSTATCOM (3334 VA) under steady-state conditions and DVR shares zero power; hence, it is not utilized completely.

Power angle affects the compensating voltage V_{DVR}. For compensation, a larger angle requires a larger amount of voltage. In a steady-state, the series transformer, on the other hand, is remaining ideal.

6.5.2 Analysis of VA Loading of UPQC with Optimization

Table 6.5 and Table 6.6 show the output results of various optimization algorithms such as PSO, TLBO, basic JAYA, SA-JAYA, SAMP- JAYA, and SAMPE-JAYA with the elite size of 4 and 14 implemented for optimum VA loading of UPQC-VA$_{min}$ for operating conditions of voltage sag and steady-state, respectively. For this, the algorithms are run for K=0.6 to 0.9 pu during voltage sag condition, and K=0.9 pu to 1.1 pu during steady-state condition with power angle range from $\delta = 0°$ to 45°. Detailed performance analysis of the UPQC-VA$_{min}$ with proposed optimization algorithms for both the operating conditions is given in the next subsections. The robustness and efficacy of various optimization algorithms in solving the optimum VA loading problem are investigated for statistical analysis indices such as best value

Table 6.4   Parameters of UPQC-VA$_{min}$ during Voltage Sag without Optimization

K	UPQC Approach	Per Phase VA Loading (VA)			V_{DVR} (V)	Angle δ (degree)
		DVR	DSTATCOM	UPQC TOTAL		
K=0.6	UPQC-P	2223	4007	6230	93	0
	UPQC-VA$_{min}$	3577	1048	4625	148	38
K=1	UPQC-P	0	3334	3334	0	0
	UPQC-VA$_{min}$	0	3333	3333	0	0

Table 6.5 Parameters of UPQC-VA$_{min}$ during Voltage Sag with Optimization

	PSO	TLBO	JAYA	SA	SAMP	SAMPE4	SAMPE14
K	0.868	0.89	0.854	0.721	0.9	0.885	0.888
Delta	29.78	27	30.8	16.352	25.84	27.7	27.35
V$_{DVR}$	114.705	104.84	118.25	85.163	100	107.35	106.1
S$_{DVR}$	1907.515	1700	1998.2	1703.917	1614.3	1750.6	1724.2
S$_{TRANS}$	2167.383	2019.7	2239.1	2110.090	1959.8	2054.2	2035.4
S$_{DSTATCOM}$	1425.819	1633	1335.3	2311.259	1719	1582.7	1609.2
SUPQC	3333.333	3333.342	3333.549	4015.176	3333.333	3333.33	3333.33

Table 6.6 Parameters of UPQC-VA$_{min}$ during Steady-State with Optimization

	PSO	TLBO	JAYA	SA	SAMP	SAMPE4	SAMPE14
K	0.89515	0.890	0.854	0.721	0.9	0.885	0.888
Delta	26.5	27	30.8	16.352	25.84	27.7	27.35
V$_{DVR}$	103	104.84	118.25	85.163	100	107.35	106.1
S$_{DVR}$	1661.5	1700	18.574	1100.475	0.054	1.106	0.813
S$_{TRANS}$	1991.7	2019.7	1998.138	1300.879	1614.295	1750.620	1724.568
S$_{DSTATCOM}$	1671.8	1633	1998.2	1703.917	1614.3	1750.6	1724.2
SUPQC	3333.334	3333.342	18.574	1100.475	0.054	1.106	0.813

(BEST), worst value (WORST), minimum value (F_{min}), mean value (F_{mean}), and standard deviation (SD). A comparison of the results of all algorithms is presented in Table 6.7 for sag and steady-state conditions, respectively. For each operating condition, independent trials of all the algorithms with different initial populations are carried out.

6.5.2.1 Analysis during Voltage Sag (0.6 pu–0.9 pu)

From the output results presented in Table 6.5, it is observed that in case of SAMP-JAYA, algorithm less voltage is required for mitigation of sag as the optimal value of power angle δ is small compared to other algorithms. VA loading of UPQC is minimum for SAMP-based approaches (3333.333VA) as compared to other optimization approaches. Figure 6.5(top) shows the VA loading of components of UPQC during voltage sag conditions. It is observed that with optimization approaches, VA loading of both converters and series transformers is reduced considerably. VA loading of the series transformer remains the same as that of DVR in case of UPQC-P; however, for UPQC-VA$_{min}$, series transformer is loaded 34% higher than the VA loading of DVR. Out of all the optimization approaches, in SAMPE-JAYA with elite size 14, the series transformer is loaded 17.34% higher than the loading of DVR, which is minimum loading among all other best approaches. Figure 6.5(bottom) shows the % utilization of the PECs for their VA rating (for an analysis, utilization rate of individual PEC is calculated by taking the ratio of VA loading of PEC to its VA rating). It is well known that the efficiency of the PEC is highest and almost constant for the load ranging from 30% to 75% of the rated capacity. From Figure 6.5(bottom), it is clear that in the case of SAMP approaches, both PECs are operated in the high-efficiency region. However, in the case of elitism-based approaches, DVR is burdened near 70% of its full load rating and DSTATCOM to 55% of its rated capacity. Convergence characteristics of proposed algorithms during voltage sag are shown in Figure 6.6(top). SAMP approaches are having a minimum function value (3333.333VA). However, SAMP-JAYA converges after the 6th iteration, and SAMPE-JAYA with elite size 4 converges after the 4th iteration whereas SAMPE-JAYA with elite size 14 converges to minimum function value after the 2nd iteration. From the statistical analysis of the output results presented in Table 6.7, it is clear that the best, worst, minimum, and average values for the SAMPE-JAYA with elite size 14 are all quite close to one another, resulting in small standard deviation values (0.000220). The value of SD confirms the robustness of the SAMPE-JAYA algorithm with elite size 14 and its ability to find the best solution in every run.

6.5.2.2 Analysis during Steady-State Condition (0.9 pu–1.1 pu)

The utilization of DVR during steady-state for load reactive power compensation minimizes the burden on DSTATCOM and the overall UPQC system as well. For

Table 6.7 Comparative Results of Different Algorithms

		PSO	TLBO	JAYA	SA	SAMP	SAMPE4	SAMPE14
Voltage Sag	BEST	--	--	3333.549	4015.176	3333.333	3333.33	3333.33
	WORST	--	--	4023.568	4013.127	3333.398	3334.374	3333.373
	Fmin	3333.333	3333.342	3333.549	4015.176	3333.333	3333.33	3333.33
	Fmean	4103.557	3333.334	3334.168	3334.128	3333.334	3333.35	3333.3333
	SD	0.002068	0.45260	0.111372	11.66	0.002743	0.010853	0.000220
Steady-State	BEST	--	--	3340.3	3338	3333.4	3333.3	3333.3
	WORST	--	--	3959.31	4030.681	3333.364	3333.43	3333.363
	Fmin	3333.5	3333.4	3340.3	3338	3333.4	3333.3	3333.3
	Fmean	3333.333	3333.333	3352.018	3353.967	3333.3334	3333.3334	3333.3333
	SD	0.056719	0.02354	0.0049568	0.4033	0.00042	0.000240	0.000231

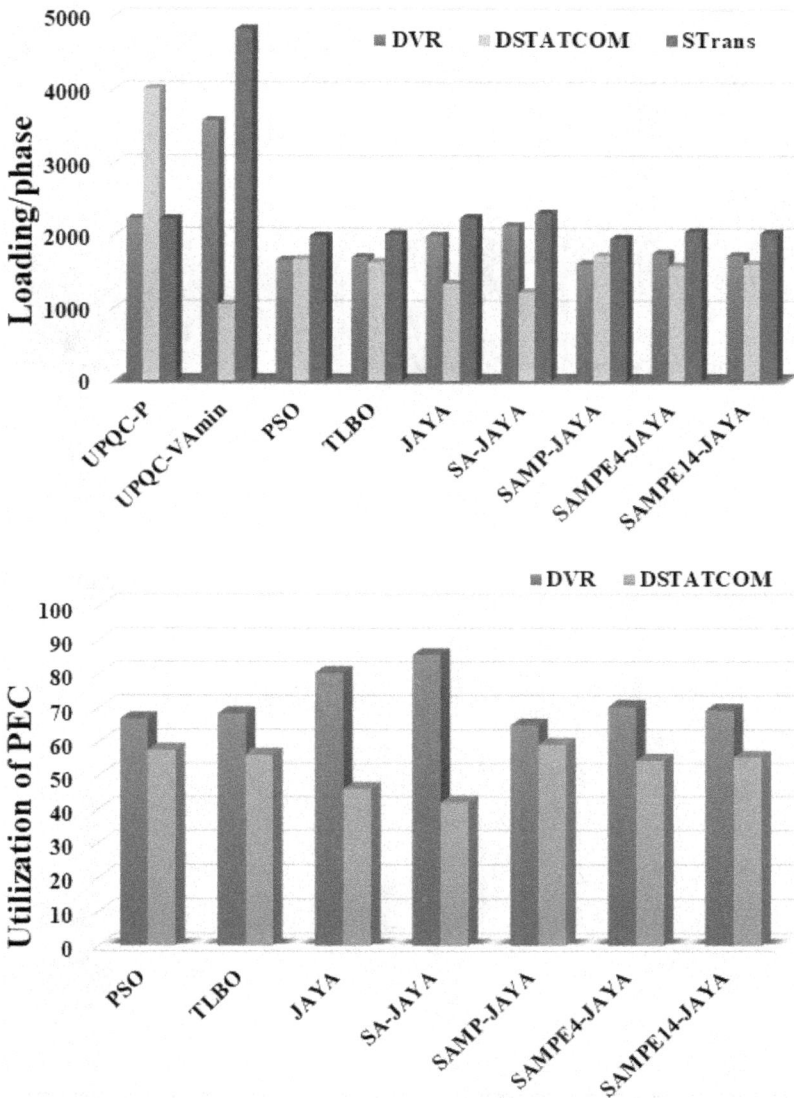

Figure 6.5 **(top) VA Loading of Individual Component of UPQC and (bottom) Utilization Rate of PECs during Voltage Sag Condition.**

the performance analysis, proposed algorithms are run for K = 0.9 pu to 1.1 pu with varying angles δ = 0° to 45°. The output results presented in Table 6.7 show that VA loading of UPQC is the least (3333.3 VA) for SAMPE-JAYA approaches compared to other optimization approaches. Figure 6.7 (top) represents the VA loading of converters of UPQC, and series transformer during a steady-state. From Figure 6.7 (top),

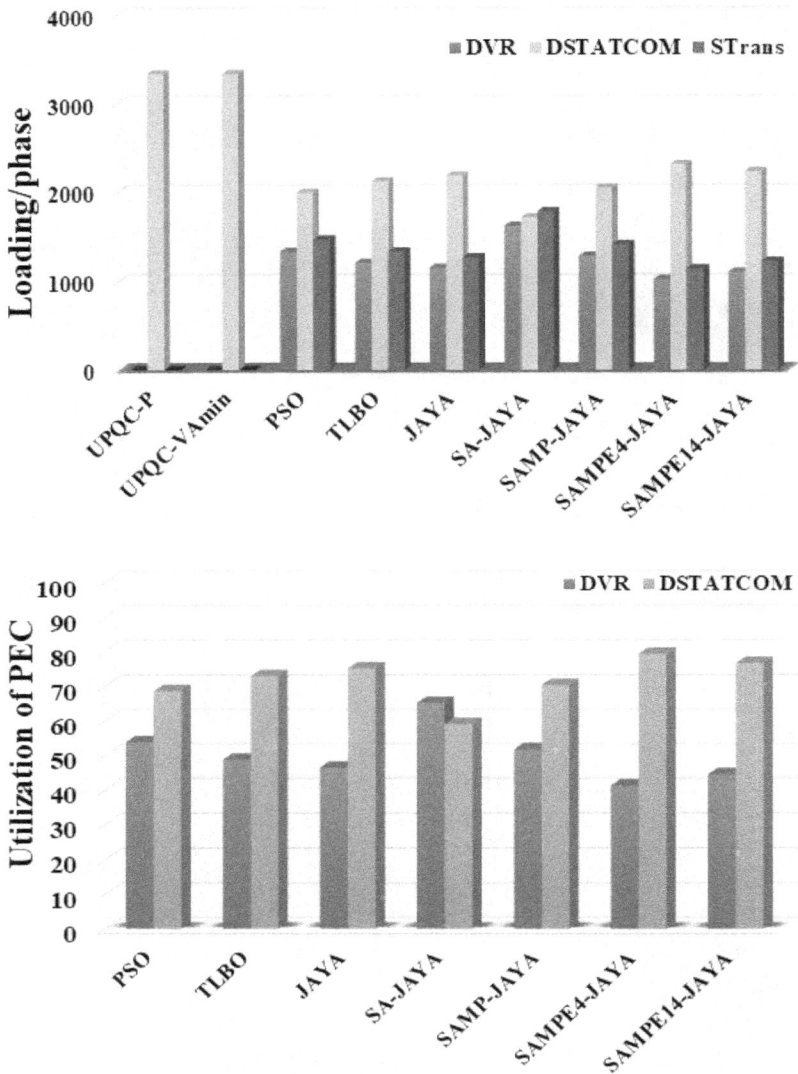

Figure 6.6 Convergence Characteristics for SAMP Approaches of JAYA Optimization Technique for the Minimal Function Value (top) for Voltage Sag Condition and (bottom) for Steady-State Condition.

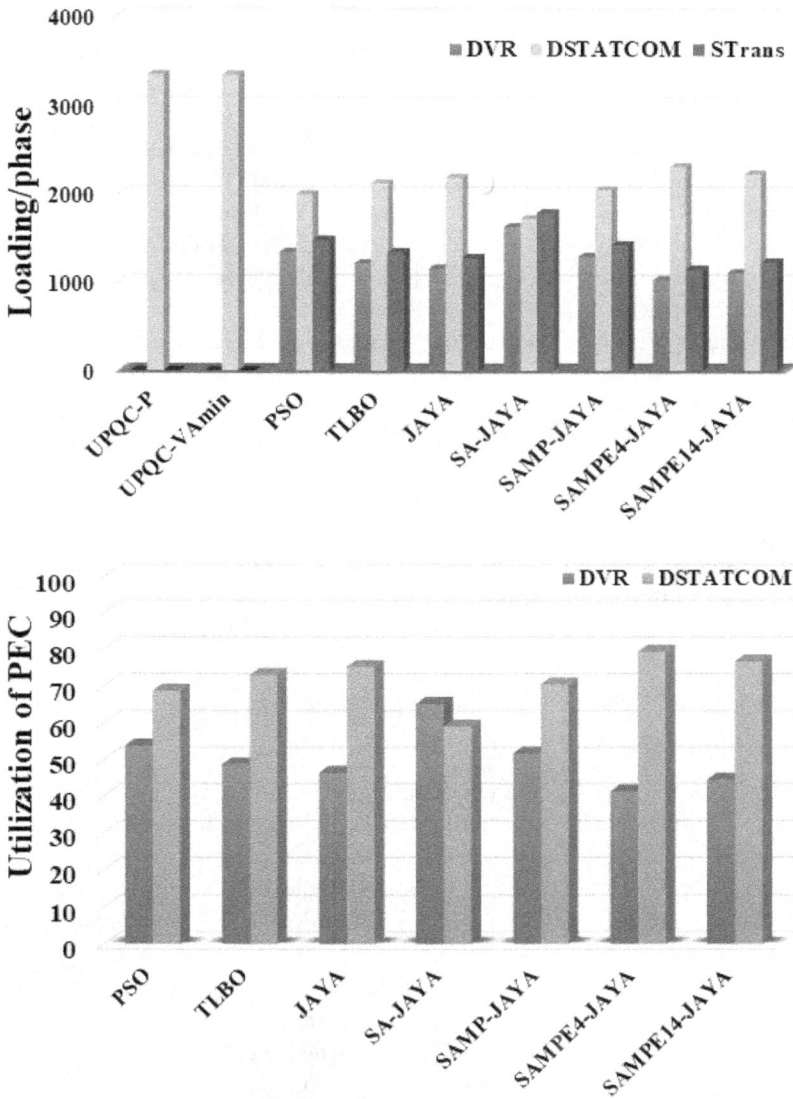

Figure 6.7 (top) VA Loading of Individual Component of UPQC and (bottom) Utilization Rate of PECs during Steady-State Condition.

it is clear that with optimization, DVR shares some amount of load reactive power, which significantly reduces the burden on DSTATCOM. It is observed that VA loading of DVR is more (1106.3 VA) in the elite size 14 approach as compared to the elite size 4 approach (1026.1VA). Although the optimal angle δ (17.11°) and series injected voltage (68 V) are the smallest for elite size 4, in the elite size 14 approach, VA loading of series transformer is minimum (10.9 % more than the VA loading of DVR). Figure 6.7 (bottom) shows the % utilization of both PECs. From Figure 6.7 (bottom), it is clear that in the case of the SAMPE- JAYA approach, VA loading of DVR is about 42% of its full load rating whereas DSTATCOM is loaded near 75% of its full load rating.

From Table 6.7, it is concluded that for steady-state conditions, SAMPE-JAYA approaches perform better than other optimization techniques. However, a statistical analysis of the results shows that the best, worst, minimum, and average values for the SAMPE-JAYA with elite size 14 are almost the same, resulting in small standard deviation values (0.000231). The robustness of the SAMPE-JAYA algorithm with elite size 14 and its ability to discover the optimum solution in every run is confirmed by the value of SD. Figure 6.6 (bottom) shows the convergence characteristics of different algorithms during a steady-state. SAMPE approaches are having a minimum function value (3333.3VA). However, SAMPE-JAYA with elite size 4 converges after the 3rd iteration, and SAMPE-JAYA with elite size 14 converges to minimum function value after the 2nd iteration.

6.5.3 Discussions

From the literature studied, different design algorithms based on fixed PAC [16] and variable PAC approach [17] have been proposed. Moreover, for determination of the most effective VA rating of individual components of the UPQC system, full load and worst conditions of sag and swell are considered. However, no optimization approach has been applied to determine the optimal VA loading of individual components of the UPQC system with a changing load, variable sag, and swell conditions.

As seen in Figure 6.3, the DVR must add a series VDVR at an angle γ with the source voltage to maintain the power angle constant. V_{DVR} and angle γ are easy to calculate because they depend on load current. The control algorithm should maintain the angle δ to retain the VA loading of both PECs below their full load ratings in any operating condition. Based on the output results of the algorithm after running several times, it is found that inequality constraints keep the VA loading of DVR and DSTATCOM and hence overall UPQC system within its full load VA rating regardless of the operating conditions.

From Table 6.4 and Table 6.5, it is concluded that during voltage sag, the burden of DVR is reduced considerably due to optimization techniques implementation as compared to without optimization approaches. As the burden of the series transformer and DVR are interdependent with optimization, it is possible to decrease the burden of the series transformer to a minimum value (around 10% higher than the

VA loading of DVR). From Table 6.4 and Table 6.7, it is concluded that compared to without optimization approaches, the VA load on DSTATCOM is reduced by around 40% in case of with optimization approaches. VA loading of all the components of the UPQC system is reduced up to a great extent after implementing the optimization algorithms. Although the UPQC system comprises all static devices such as PECs and series transformers, the high switching frequency of PECs causes higher losses than a series transformer. As a result, the efficacy of the UPQC is mainly dependent on the losses that take place in both converters. Higher VA loading of UPQC for correction of the equal amount of voltage sag/swell leads to larger operating power losses. As a result of the application of the optimization, the VA loading of DVR and DSTATCOM is reduced while the compensating capabilities are maintained, resulting in reduced operating power losses and increased efficiency. Figure 6.6 (top) and Figure 6.6 (bottom) are convergence graphs. That illustrates that the SAMPE-JAYA algorithm with elite size 14 has excellent convergence properties compared to other algorithms for solving the optimal VA loading of the UPQC problem.

6.6 Conclusion

In this chapter, an intelligent algorithm based on SAMPE-JAYA is proposed for determining the optimum angle and K for minimizing the UPQC VA burden. The results of the proposed algorithm are compared with PSO, TLBO, basic JAYA, and its various variations to show its usefulness. Based on the output findings, it can be concluded that SAMPE-JAYA performs better than other algorithms. The advantages of the proposed algorithm are as follows: 1. It is based on the variable PAC method; 2. It doesn't require algorithm-specific parameters; 3. Multi-population concept improves the diversity of search; and 4. Replacement of worst solutions with elite solutions accelerates the convergence. Optimum utilization of both PECs will result in power loss and manufacturing cost reduction. This makes the UPQC system more appealing from a cost perspective, as well as an effective option for PQ issue mitigation and DG grid integration for clean and green energy. As an outcome of this research, a control approach for UPQC based on optimum instantaneous VA loading will be developed to optimize the efficacy and VA loading of UPQC.

Symbols

I_{source}, V_{source}	Source current and voltage during normal condition
I'_{source}, V'_{source}	Source current and voltage during abnormal condition
$I_{DSTATCOM}$	Current injected by DSTATCOM
$P_{DSTATCOM}, P_{DVR}$	Active power handled by DSTATCOM and DVR, respectively
$Q_{DSTATCOM}, Q_{DVR}$	Reactive power handled by DSTATCOM and DVR, respectively
$S_{DSTATCOM}, S_{DVR}$	Burden on DSTSTACOM and DVR, respectively

$S_{TOTALUPQC}$	Burden on UPQC
S_{TRANS}	Series transformer VA loading
V_{DVR}	Voltage added in series by DVR
V_{Load}, I_{Load}	Load end voltage and current
δ	Swing/power angle
γ	Angle between V_{source} and V_{DVR}

References

1. R. C. Dugan, *Electic Power System Quality*. McGraw-Hill, 2004.
2. Math H. J. Bollen, *Understanding Power Quality Problems Voltage Sag, and Interruption*. Wiley-IEEE Press, 2000.
3. Bhim Singh, *Power Quality Problems and Mitigation Techniques*. John Wiley & Sons, Ltd, 2015.
4. A. Ghosh and G. Ledwich, *Power Quality Enhancement Using Custom Power Devices*. Kluwer, 2002.
5. H. Fujita and H. Akagi, "The unified power quality conditioner: The integration of series- and shunt active filters," *IEEE Trans. Power Electron.*, vol. 13, no. 2, pp. 315–322, 1998, DOI: 10.1109/63.662847.
6. M. T. L. Gayatri and A. M. Parimi, "Mitigation of supply & load side disturbances in an AC Microgrid using UPQC," in *2016 IEEE 6th International Conference on Power Systems, ICPS 2016*, 2016, DOI: 10.1109/ICPES.2016.7584116.
7. V. Khadkikar, "Enhancing electric power quality using UPQC: A comprehensive overview," *IEEE Trans. Power Electron.*, vol. 27, no. 5. pp. 2284–2297, 2012, DOI: 10.1109/TPEL.2011.2172001.
8. S. Gade, R. Agrawal, and R. Munje, "Recent trends in power quality improvement: Review of the unified power quality conditioner," *ECTI Trans. Electric. Eng. Electron. Commun.*, vol. 19, no. 3, pp. 268–288, 2021. https://doi.org/10.37936/ecti-eec.2021193.244936.
9. W. C. Lee, D. M. Lee, and T. K. Lee, "New control scheme for a unified power-quality compensator-Q with minimum active power injection," *IEEE Trans. Power Deliv.*, vol. 25, no. 2, pp. 1068–1076, 2010.
10. V. Khadkikar and A. Chandra, "UPQC-S: A novel concept of simultaneous voltage sag/swell and load reactive power compensations utilising series inverter of UPQC," *IEEE Trans. Power Electron.*, vol. 26, no. 9, pp. 2414–2425, 2011, DOI: 10.1109/TPEL.2011.2106222.
11. Y. Y. Kolhatkar, R. R. Errabelli, and S. P. Das, "A sliding mode controller based optimum UPQC with minimum VA loading," in *2005 IEEE Power Engineering Society General Meeting*, 2005, vol. 1, pp. 871–875, DOI: 10.1109/pes.2005.1489536.
12. Y. Y. Kolhatkar and S. P. Das, "Experimental investigation of a single-phase UPQC with minimum VA loading," *IEEE Trans. Power Deliv.*, vol. 22, no. 1, pp. 373–380, 2007, DOI: 10.1109/TPWRD.2006.881471.
13. M. Basu, M. Farrell, M. F. Conlon, K. Gaughan, and E. Coyle, "Optimal control strategy of UPQC for minimum operational losses," in *39th International Universities Power Engineering Conference, UPEC 2004—Conference Proceedings*, 2004, vol. 1, pp. 246–250.

14. S. Dutta, P. K. Roy, and D. Nandi, "Optimal location of UPFC controller in transmission network using hybrid chemical reaction optimisation algorithm," *Int. J. Electr. Power Energy Syst.*, vol. 64, pp. 194–211, 2015, DOI: 10.1016/j.ijepes.2014.07.038.
15. G. Siva Kumar, B. Kalyan Kumar, and M. Mahesh Kumar, "Optimal VA loading of UPQC during mitigation of unbalanced voltage sags with phase jumps in three-phase four-wire distribution system," in *2010 International Conference on Power System Technology: Technological Innovations Making Power Grid Smarter, POWERCON2010*, 2010, pp. 1–8, DOI: 10.1109/POWERCON.2010.5666492.
16. B. B. Ambati and V. Khadkikar, "Optimal sizing of UPQC considering VA loading and maximum utilisation of power-electronic converters," *IEEE Trans. Power Deliv.*, vol. 29, no. 3, pp. 1490–1498, 2014, DOI: 10.1109/TPWRD.2013.2295857.
17. J. Ye, H. B. Gooi, and F. Wu, "Optimal design and control implementation of UPQC based on variable phase angle control method," *IEEE Trans. Ind. Informatics*, 2018, DOI: 10.1109/TII.2018.2834628.
18. V. Khadkikar, "Fixed and variable power angle control methods for unified power quality conditioner: Operation, control and impact assessment on shunt and series inverter kVA loadings," *IET Power Electron.*, 2013, DOI: 10.1049/iet-pel.2012.0715.
19. J. Ye, H. B. Gooi, X. Zhang, B. Wang, and U. Manandhar, "Two-level algorithm for UPQC considering power electronic converters and transformers," in *Conference Proceedings—IEEE Applied Power Electronics Conference and Exposition—APEC*, 2019, vol. 2019–March, pp. 3461–3467, DOI: 10.1109/APEC.2019.8722007.
20. G. S. Kumar, P. H. Vardhana, B. K. Kumar, and M. K. Mishra, "Minimisation of VA loading of unified power quality conditioner (UPQC)," in *POWERENG 2009–2nd International Conference on Power Engineering, Energy and Electrical Drives Proceedings*, 2009, DOI: 10.1109/POWERENG.2009.4915152.
21. D. O. Kisck, V. Navrapescu, and M. Kisck, "Single-phase unified power quality conditioner with optimum voltage angle injection for minimum VA requirement," in *IEEE International Symposium on Industrial Electronics*, 2007, DOI: 10.1109/ISIE.2007.4374990.
22. Swati Gade and Rahul Agrawal, "Optimal utilisation of UPQC during voltage sag and steady-state using particle swarm optimization," Intelligent Computing in Information Technology for Engineering System, 2021, ISBN: 978-1-032-27080-7.
23. S. Gade, R. Agrawal, D. Patil, and S. Antonov, "Optimal Utilisation of UPQC at Different Operating Condition Using TLBO," in *2021 56th International Scientific Conference on Information, Communication and Energy Systems and Technologies (ICEST)*, 2021, pp. 197–200, DOI: 10.1109/ICEST52640.2021.9483567.
24. Swati Gade and Rahul Agrawal, "Optimal utilisation of unified power quality conditioner using the JAYA optimisation algorithm," *Engineering Optimization*, 2021, DOI: 10.1080/0305215X.2021.1978440.
25. S. A. Gade and R. Agrawal, "Optimal utilisation of UPQC during steady-state using evolutionary optimisation techniques," in *2021 International Conference on Intelligent Technologies (CONIT)*, 2021, pp. 1–7, DOI: 10.1109/CONIT51480.2021.9498479.
26. A. Gotmare, S. S. Bhattacharjee, R. Patidar, and N. V. George, "Swarm and evolutionary computing algorithms for system identification and filter design: A comprehensive review," *Swarm Evol. Comput.*, vol. 32, pp. 68–84, 2017, DOI: 10.1016/j.swevo.2016.06.007.
27. R. Venkata Rao, *Teaching Learning-Based Optimisation Algorithm: And Its Engineering Applications*. Springer Publication, 2015.

28. R. Venkata Rao, "Review of applications of turbo algorithm and a tutorial for beginners to solve the unconstrained and constrained optimisation problems," *Decis. Sci. Lett.*, vol. 5, no. 1, pp. 1–30, 2016, DOI: 10.5267/j.dsl.2015.9.003.

29. R. Venkata Rao, "Jaya: A simple and new optimisation algorithm for solving constrained and unconstrained optimisation problems," *Int. J. Ind. Eng. Comput.*, vol. 7, no. 1, pp. 19–34, 2016, DOI: 10.5267/j.ijiec.2015.8.004.

30. R. Venkata Rao and A. Saroj, "A self-adaptive multi-population based Jaya algorithm for engineering optimisation," *Swarm Evol. Comput.*, 2017, DOI: 10.1016/j.swevo.2017.04.008.

31. C. Li, T. T. Nguyen, M. Yang, S. Yang, and S. Zeng, "Multi-population methods in unconstrained continuous dynamic environments: The challenges," *Inf. Sci. (NY).*, vol. 296, no. 1, pp. 95–118, 2015, DOI: 10.1016/j.ins.2014.10.062.

32. R. V. Rao and A. Saroj, "An elitism-based self-adaptive multi-population Jaya algorithm and its applications," *Soft Comput.*, vol. 23, no. 12, pp. 4383–4406, 2019, DOI: 10.1007/s00500-018-3095-z.

33. R. V. Rao and G. G. Waghmare, "A new optimisation algorithm for solving complex constrained design optimisation problems," *Eng. Optim.*, 2017, DOI: 10.1080/0305215X.2016.1164855.

34. R. V. Rao and K. C. More, "Design optimisation and analysis of selected thermal devices using self-adaptive Jaya algorithm," *Energy Convers. Manag.*, vol. 140, pp. 24–35, 2017, DOI: 10.1016/j.enconman.2017.02.068.

Chapter 7

Applications of Artificial Intelligence

Anuprita Mishra and Anita Soni

Contents

DOI: 10.1201/9781003301820-7

7.1 Introduction: AI Techniques

Artificial intelligence is defined as "Intelligent machine or software developed by software expert or developer." It is a way of making computer-oriented robotics, or a software that thinks intelligently, like a human being. Artificial intelligence is a young technology, but its value is apparent.

Today's artificial intelligence plays a wide role in every field of science and technology, specially computer science, biology, psychology, linguistics, mathematics, and engineering. A major thrust of AI is in the development of computer functions associated with human intelligence, such as reasoning, learning, and problem solving [1].

AI provides generic solutions to any real-world problems. It helps us in modifying data without doing any changes. By using accumulated information of machine, using AI, human being are highly dependent, it divides data into three different tasks [2]. See Figure 7.1 and Figure 7.2.

7.2 Applications of AI

- Playing games
- Understanding natural language processing
- Expert systems
- Understanding speech
- Handwriting recognition

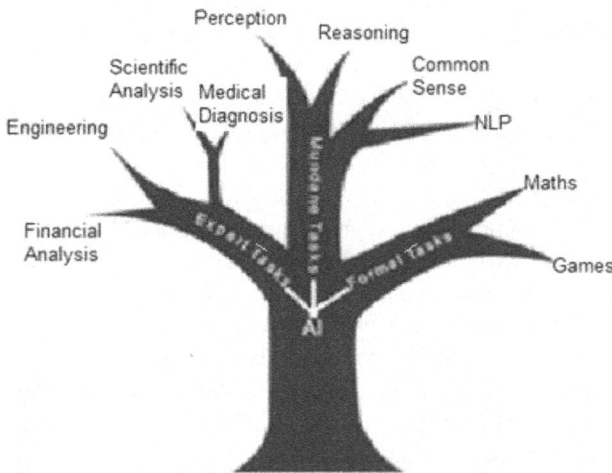

Figure 7.1 Mentioning Different Tasks.

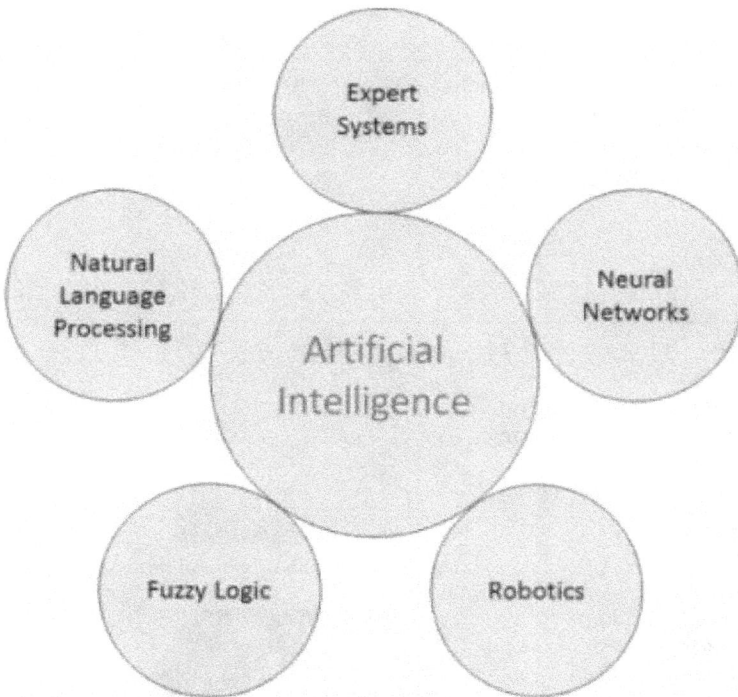

Figure 7.2   Domains of Artificial Intelligence.

- ■ Vision systems
- ■ Intelligent robots

While talking about AI, we must understand the term intelligence and knowledge, which refer to the ability of a system to accumulate reason; perceive relationships and analogies; learn from experience; store and retrieve information from memory; solve problems; comprehend complex ideas; and based on natural language, classify, generalize, and adapt in new situations. [3]

7.3 Different Modern Techniques

Which played very important role in power system.

1. ANNs (artificial neural networks)
2. FL (fuzzy logic systems)
3. XPS (expert system techniques)
4. GA (genetic algorithm)

Example—Working with the help of a computer system such as booking tickets, tracking flight details, clinical world, developer	 Figure 7.3a—Expert system
Example—Doing any work by Google Translator, Google Feature, speech recognition, automatic voice recognition	 Figure 7.3b—Natural language processing
Example—Helps in pattern recognition, face recognition, character recognition, handwriting recognition	 Figure 7.3c—Neural networks
Example—Helps in handling industrial works more efficiently: Spraying, Painting, Coating, Carving	 Figure 7.3d—Robotics
Example—Fuzzy logic system is used as inputs degree of trust and linguistic variable. Consumer electronics, automobiles	 Figure 7.3e—Fuzzy logic

Figure 7.3   Different Uses of AI.

7.3.1 Artificial Neural Networks

Dr. Robert Hecht-Nielsen defines artificial neural network (ANN) as a computing system made up of a number of simple, highly interconnected processing elements, which exchange information by change in external inputs in their dynamic state [3].

ANN is an inspired system that works on primary advantages based algorithm and simulates online adaption of dynamic systems, quick parallel computation, and intelligent interpolation. It's architecture based on several layers, topology, connectivity pattern, feed forward, back propagation and radial function or recurrent, etc. Each neural [4] network processing information concludes a function from approximation, classification, and data processing.

7.3.1.1 ANN's Structure

The human brain is composed of 86 billion nerve cells called neurons. They are connected by using silicon and wires as living neurons and dendrites to other thousand cells by axons [5]. ANN is based on the right connection of the human brain. Inputs accepted by external environment are sensory organs called dendrites. A neural network sends quick inputs message to other neurons for handling issue, created by electric impulses [6]. See Figure 7.4.

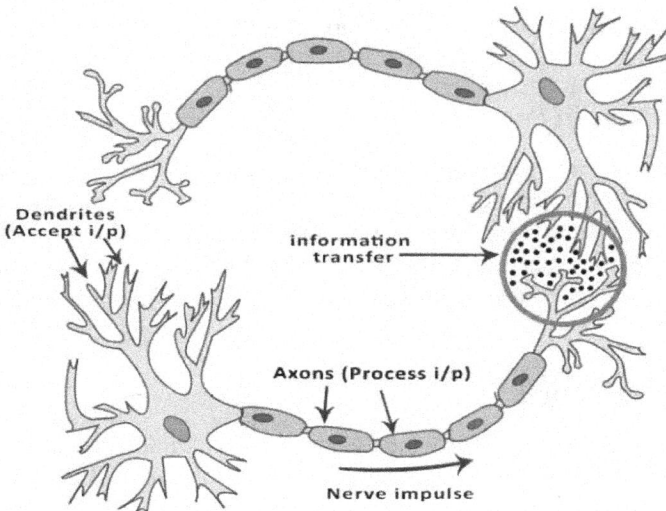

Figure 7.4 Basic Structure of ANNs.

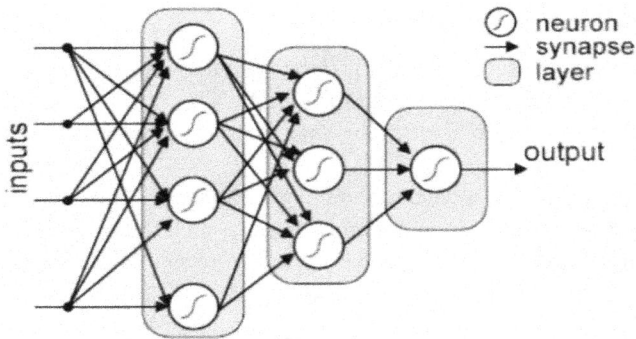

Figure 7.5   Hidden Layers.

ANNs resemble biological neurons of the human brain, which includes multiple nodes. The neurons accept input data, and they interact with each other in performing simple operations on data for giving activation [7] or node value.

Artificial neurons have layers (see Figure 7.5):

- **Input Layer:** The input units help us in distributing data to other nodes but do not process any output value.
- **Hidden Layer:** The hidden units help us to solve any complex nonlinear problems.
- **Output Layer:** Considering the output result helps us to encode possible values.

7.3.1.2 ANNs in Power Systems

Any basic problems (generation, transmission, and distribution) of electricity will get solved. ANN is fast and robust [8–10]. It has a fault tolerant tendency that does not need any appropriate knowledge model. By entering the exact consumption parameter of particular area (transmission and distribution) by that we are able to determine the disturbance in power supply.

7.3.2 Fuzzy Logic

Fuzzy systems developed in 1965 and used in. solving technical problems. It helps us in making decisions based on standardized [11–13] and systematized approximate reasoning, describing ambiguity in linguistic terms instead of exact mathematical description. It produces accurate solutions based on approximate information and data.

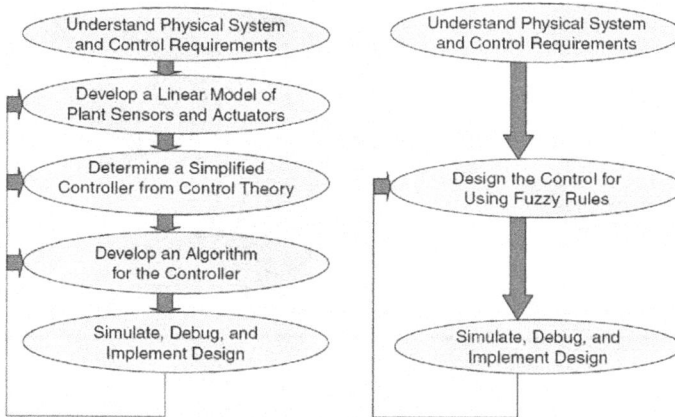

Figure 7.6   Benefits of Fuzzy Logic.

7.3.2.1 Fuzzy Logic Characteristics

Fuzzification solves complex problem in low cost by applying the oversimplification method of superior expressive power, and an improved capability [14–16]. See Figure 7.6.

Fuzzy logic is used for solving small or complex problems related to hardware or software mode [17]:

1. Enhancement stability analysis
2. Power controls system
3. Fault tolerance and fast diagnosis
4. Full security assessment
5. Evaluate load forecasting
6. Active power planning and control
7. Evaluate state estimation

7.3.2.2 Fuzzy Logic Controller

According to requirement, it can be used for small circuits to large mainframes, that is, software or hardware. Adaptive fuzzy. [18] controllers help to control.complex process. See Figure 7.7.

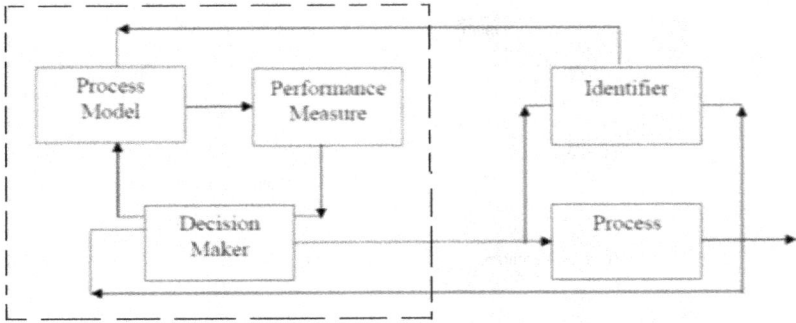

Figure 7.7 Model-Based Controller.

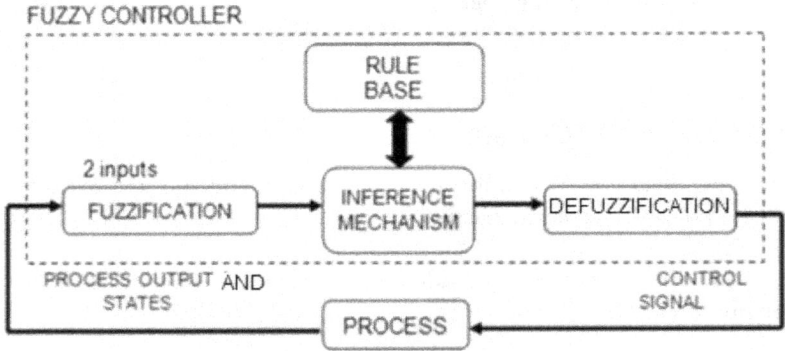

Figure 7.8 Components of Fuzzy Logic Controller.

7.3.2.3 Enhancing the Voltage Profile of Power System

1. Enhancing voltage deviation and controlling ability of device based on demand and [19] control facilities.
2. Voltage changes and controlling variables are converted (fuzzy set).
3. Based on fuzzy set notation, the operation of existing.voltage and control device are arranged as in Figure 7.8.

7.3.2.4 Fuzzy Logic in Power Systems

Fuzzy logic is used to increase the efficiency of physical components of power systems by using computing-based "degrees of truth" rather than the usual "True or False" (0 or 1) Boolean Logic [20].

7.3.3 Expert Systems

Expert systems are also known as knowledge or rule-based systems, developed during the 1960s and 1970s and commercially applied throughout from the 1980s. Expert systems collect data based on information and past experience, which are combined together as knowledge and interface mechanisms [21]. It uses knowledge and information mechanism to solve problems, which is difficult to solve intellect by any human being, where AI help us to organize data and facts about the task domain. The success of any expert system [22, 23] depends upon the collection of highly accurate and precise knowledge.

Knowledge base of expert system consists of the following (see Figure 7.9):

- **Factual Knowledge**: Information that is widely accepted by the expert knowledge engineers and scholars.
- **Heuristic Knowledge**: It is about practice, accurate judgment, one's ability to evaluate, and guessing.

7.3.3.1 Knowledge Representation

Expert systems have the ability to solve problems related to any field, which helps others to organize and formalize knowledge base, which stores knowledge [24] separately from the program's procedural part to solve a problem of any form. See Figure 7.10.

An expert system provides permanent, consistent, and easy documented solution based on any method neural networks, object-oriented methodology, case-based reasoning, system architecture, intelligent agent systems, database methodology,

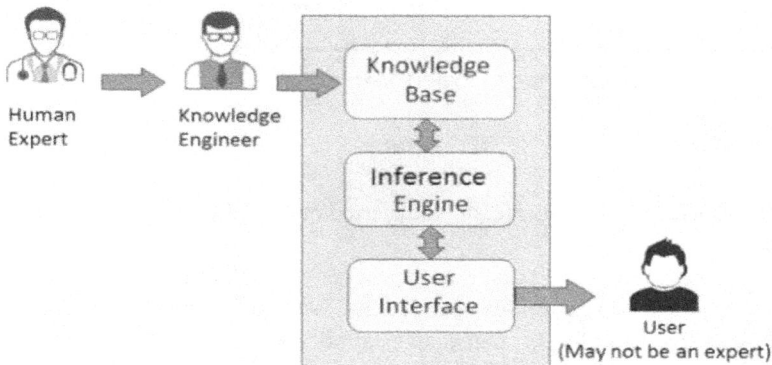

Figure 7.9   Components of KBS.

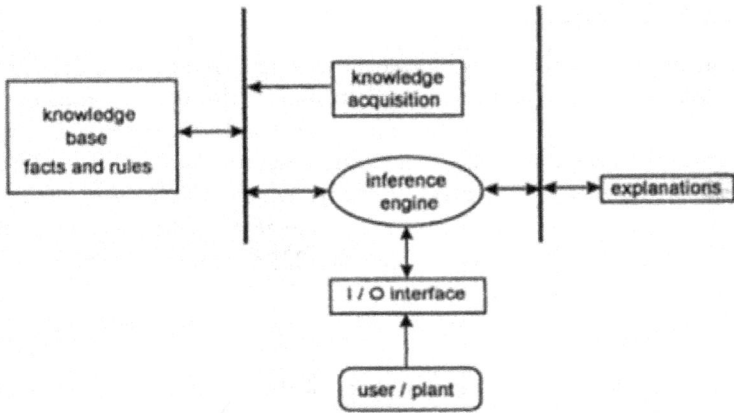

Figure 7.10   Structure of a Knowledge Base Expert System.

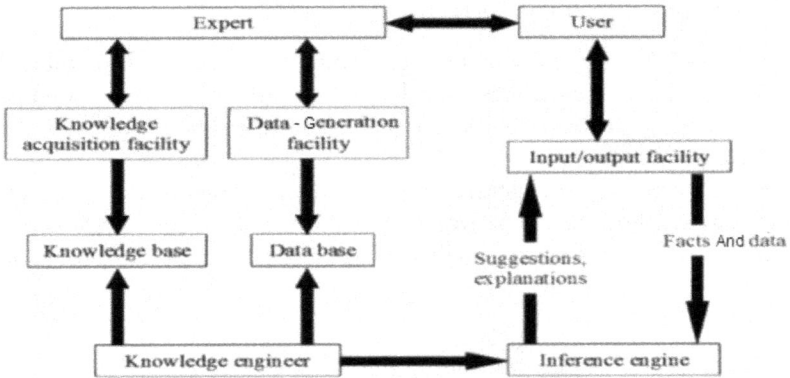

Figure 7.11   Structure of the AI Expert System.

modeling, and ontology, which easily transferred or reproduced [25] knowledge-based systems.

Knowledge based systems help in expert system evaluation (Figure 7.11) which depends on fuzzy and neuron-expert systems. Worked in creating Expert advanced computer techniques [26–28] that may be applicable in the online problem-solving mode of the power system.

7.3.3.2 Expert Systems in Power Systems

Intelligent systems or autonomous expert systems in monitoring instruments help engineers to detect problem (system condition) rapidly. This is especially important when restoring service by major disturbances. Expert system works on different module by monitoring fault; it generates information rather than just collect data. Virtually, estimation of these values can be done and further research for increasing the efficiency of the process can also be performed [29]. We are easily modifying design (Figure 7.12) due to computer programming.

Different modules capture power quality, and each module invokes unique knowledge that is available to the user by sending email, fax, or message, which play an important role in power management. See Figure 7.12.

Artificial Intelligent Techniques Over Traditional Methods

A.   Advantages

■ Artificial intelligent methods are applicable to smart grid because of its modernity.

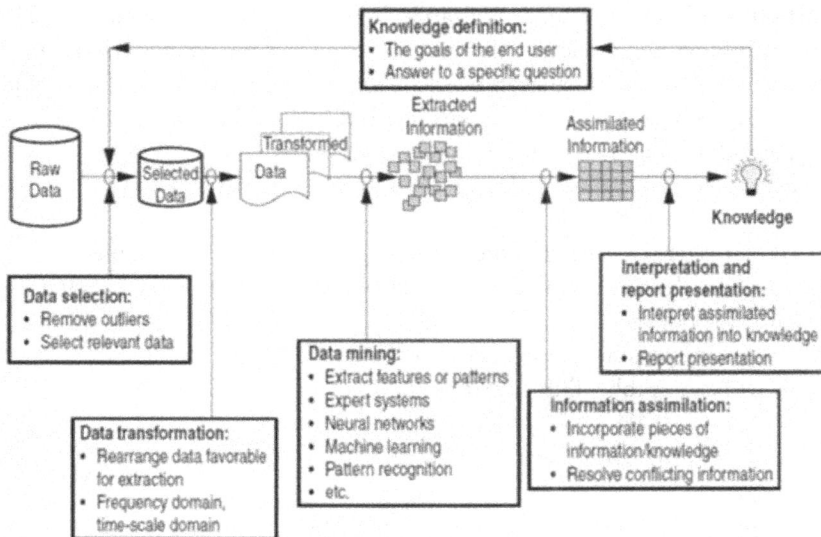

Figure 7.12   Steps in Turning Raw Data into Expert Knowledge.

- Genetic algorithm needs only rough information of the objective function and places no restriction such as.differentiability and convexity on the objective function.
- Genetic algorithm works with a set of solutions from one generation to the next, and not a [30] single solution. (Converge on local minima.)
- In the genetic algorithm, the solutions are randomly developed based on the probability rate of the genetic operators, such as mutation and crossover.
- Fuzzy logic more accurately represents the operational constraints of power systems, and fuzzified constraints are softer than traditional constraints.
- The advantages of simulated annealing are general applicability in dealing with arbitrary systems and cost functions, its [31] ability to refine optimal solution, and its simplicity of implementation even for complex problems.

B.   Disadvantages

- Poor computational search is the main drawback of above technique.
- The major drawback is repeating simulated annealing.
- Genetic algorithm method requires [32] tremendously long time.

7.4 Genetic Algorithms

Genetic algorithm (GA) is an optimization technique based on the study of "Natural selection and natural Genetics." It's excitation to solve the voltage control problem and reactive power compensation by increasing the efficiency and analysis of power system output. Genetic algorithm withstands all selected constraints [33]. It is the best method for solving complex and nonlinear problems.

Genetic algorithm differentiates from optimization methods:

1. It works on the coding of the variables set, not on actual variables.
2. Genetic algorithm evaluates optimal points of any problem by checking all possible points.
3. It uses information for generating objective function.
4. It works on probability transition.

It has the following components:

1. Initial problems of lines calculate individually.
2. An evaluation function plays ranking the individuals in terms of power productivity and transmission.
3. Genetic algorithm determines the configuration of a new problem generated based on previous solution [34].

7.4.1 GA Applications in Power System

Genetic algorithms are based on the principle of survival of the fittest.methods, which help us in increasing the efficiency of power system processes and increasing power output. Problems focused widely on power systems cannot be solved by conventional techniques. [35, 36].Some of the areas of the power system applications are as follows:

1. Planning—Proper planning in distribution of electricity:
   - Positioning of wind turbine
   - Active optimization of power
   - Feed the routing network
   - By capacitor placement
   - Economic load dispatch, generation and operational planning based on load forecasting, optimization hydrothermal generation scheduling
   - Power transmission capacity and optimal power flow helps in real and reactive. power limits of generators and system reliability

2. Operation—Help in maintenance and scheduling power plant.
   - Hydrothermal plant.coordination
   - Proper scheduling of maintenance
   - Minimization of the loss of.power
   - Help in managing load
   - Control of FACTS
   - Control of voltage and frequency for system stability, sizing, and control of FACTS devices
   - Analysis of the electricity markets and strategies for bidding

3. Analysis—Help in analysis of load shading and flow
   - Harmonic distortion reduction
   - Design filtration
   - Balance in frequency control and maintained load flow
   - Automation power restoration and manage security margins easily, which helps in fault diagnosis

7.4.2 AI Techniques for Transmission Line Performance Improvement

A. practical application to improve the performance of transmission line is described with the help of a combination of AI [37] techniques.

- **Fuzzy systems**: To diagnose the fault.
- **ANNs**: Trained to change the values of line parameters based on environmental conditions.

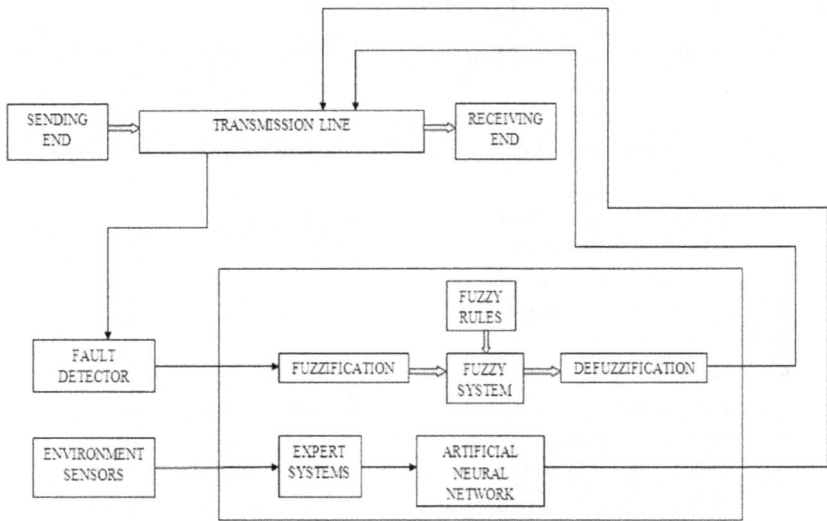

Figure 7.13   Practical Application of AI in Power Transmission Line.

- **Expert systems:** To deploy outputs as a value of line parameters.
- **Environmental sensors**: To sense the environmental and atmospheric conditions and provide input to the expert systems.

Fuzzy systems are generally used for fault diagnosis. Only three line currents are sufficient to. implement this angular difference technique between fault and pre-fault current phases [38, 39]. If any fault occurs in transmission line, fuzzy system is used to detect fault type. See Figure 7.13.

Artificial neural networks and expert systems are used to improve the performance of the line. The environmental and atmospheric conditions sense input from the expert systems [40, 41]. The ANNs evaluate output based on environmental conditions; processing speed is directly proportional to the number of neurons. These networks take different neurons for different layers and different activation functions.

7.4.3 Observations/Outputs

- Work with hybrid artificial intelligent techniques for better performance of all optimization problems.
- Swarm intelligence has more potential.in power system analysis, and it is also the most recent in the field of computational.intelligence technique.
- Researchers indicate that simulated annealing is selected for arbitrary system, cost functions, refined optimal solution, and simplicity of implementation complex problem.

■ Artificial intelligence problems require use of knowledge bases to store human knowledge, operator judgment particularly in practical solutions, and experience gained over a period, characterization by network uncertainty, load variations,.etc.

7.4.4 Summary

Efficiency issues are particularly problematic.as the prevalence of informal connections to the power grid means a large amount of power is neither measured nor billed, resulting in losses as well as greater CO_2 emissions. The energy sector worldwide faces growing challenges related to rising demand, efficiency, changing supply and demand patterns, and a lack of analytics needed for optimal management. These challenges are more acute in emerging market nations. The power sector in developed nations has already begun to use artificial intelligence and related technologies that allow for communication between smart grids, smart meters, and Internet of Things. These technologies can help improve power management, efficiency, and transparency and increase the use of renewable energy sources. Power system optimization is aimed at improvements in more areas than cost: reliability, efficiency, economics, environmental friendliness, and security.

References

1. International Journal of Engineering Intelligent Systems. 1997. The special issue on AI Applications to Power System Protection, edited by M. M. Saha and B. Kasztenny, Vol. 5, No. 4, pp. 185–193, December.
2. Dahhaghchi, I., and R. D. Christie. 1997. "AI Application Areas in Power Systems." *IEEE Expert*, Vol. 12, Issue 1, pages 58–66, January/February.
3. Bachmann, B., D. Novosel, D. Hart, Y. Hu, and M. M. Saha. 1996. "Application of Artificial Neural Networks for Series Compensated Line Protection." Proceedings of the International Conference on Intelligent System Application to Power Systems, Orlando, January 28—February 2, pp. 68–73.
4. Anis Ibrahim, W. R., and M. M. Morcos. 2002. "Artificial Intelligence and Advanced Mathematical Tools for Power Quality Applications: A Survey." *Power Delivery, IEEE Transactions*, Vol. 17, Issue 2, Pages 668–673, April.
5. Khedher, M. Z. 1997. "Fuzzy Logic in Power Engineering." Regional Conference of CIGRE committees in Arab Countries, May 25–27, Doha, Qatar.
6. Warwick, K., A. Ekwue, and R. Aggarwal (eds). 1997. *Artificial Intelligence Techniques in Power Systems*. The Institution of Electrical Engineers, London.
7. Momoh, James A., and E. EL-Hawary Mohamed. 2000. "Electric Systems, Dynamics, and Stability with Artificial Intelligence, Marcel Dekker, Inc. USA. Smartgrid.gov. (n.d.)." *The Smart Grid*. www.smartgrid.gov/the_smart_grid/smart_grid.html.
8. Chandra, Harsh. 2019. "Artificial Intelligence (AI) vs Machine Learning (ML) vs Big Data." May 10, 2019. https://heartbeat.fritz.ai/artificialintelligence-ai-vs-machine-learning-ml-vs-big-data-909906eb6a92.

9. World Bank. 2020. "Access to Electricity (% of Population)." *Indicators.* https://data.worldbank.org/indicator/EG.ELC.ACCS.ZS.
10. "The Role of Artificial Intelligence in Achieving the Sustainable Development Goals." https://arxiv.org/ftp/arxiv/papers/1905/1905.00501.
11. Kosko, B. 1992. *Neural Networks and Fuzzy Systems.* Prentice Hall, Englewood Cliffs, NJ.
12. El-Hawary, Mohamed E. 1998. *Electric Power Applications of Fuzzy Systems.* John Wiley, IEEE press, pp. 1–37, 1998.
13. Alander, J. T. 1996. "An Indexed Bibliography of Genetic Algorithm in Power Engineering." Power Report Series 1–94.
14. Kirkpatrick S., C. D. Gelatt, and M. P. Vecchi. 1983. "Optimization by Simulated Annealing." *Science.* New Series 220, pp. 671–680. Lai, Loi Lei, 1998. *Intelligent System Applications in Power Engineering: Evolutionary Programming and Neural Networks.* John Willey & Sons, UK. Proceedings Papers.
15. Gagan, Olivia. 2018. "Here's How AI Fits into the Future of Energy." www.weforum.com, May 25, 2018. www.weforum.org/agenda/2018/05/how-ai-can-help-meet-global-energy-demand.
16. Zhang, Yang, Tao Huang and Ettore Francesco Bompard. 2018. "Big Data Analytics in Smart Grids: A Review." *Energy Inform,* Vol. 1, No. 8. https://doi.org/10.1186/s42162-018-0007-5.
17. Zhang Zhen. 2011. "Smart Grid in America and Europe: Similar Desires, Different Approaches." *Public Utilities Fortnightly,* Vol. 149, No. 1, January 1, 2011. https://papers.ssrn.com/sol3/papers.cfm?abstract_id=1799705.
18. Sagiroglu Serif, Ramazan Terzi, Yavus Canbay and Ilhami Colak. 2016. "Big Data Issues in Smart Grid Systems." 2016 IEEE International Conference on Renewable Energy Research and Applications (ICRERA), Birmingham, pp. 20–23.
19. Witherspoon, Sims and Will Fadrhonc. 2019. "Machine Learning Can Boost the Value of Wind Energy." 25-Feb-2022 — [216] Carl Elkin.
20. Sentient Energy. 2017. "Manitoba Hydro Selects Sentient Energy 4G LTE Enabled Cellular Grid Analytics to Drive Success of 'Worst Feeder Program'." www.sentient-energy.com/news/manitoba-hydro-selects-sentient-energy-4g-lte-enabled-cellular-grid-analytics.
21. Wehus, Walter Norman. 2017. "Soon, Artificial Intelligence Can Operate Hydropower." *University of Agder,* January 20, 2017.
22. OECD/IEA. 2018. "Electricity Access Database." www.iea.org/sdg/electricity/.
23. Vinuesa, Ricardo, Hossein Azizpour Hossein, Iolanda Leite, Madeline Balaam, Virginia Dignum, Sami Domisch, Anna Felländer, Simone Langhans, Max Tegmark, and Francesco Fuso Nerini. "The role of artificial intelligence in achieving the Sustainable Development Goals", *Natural Communication,* pp. 1–10, 2020.
24. Much of this paragraph was informed by Rastgoufard, Samin. 2018. "Applications of Artificial Intelligence in Power Systems." University of New Orleans Theses and Dissertation, No. 2487. May 18, 2018. https://scholarworks.uno.edu/td/2487.
25. Thakker, Aman Y. 2019. "By the Numbers: India's Progress on its Renewable Energy Target." *CSIS cogitAsia,* February 27, 2019. www.cogitasia.com/by-the-numbers-indias-progress-on-its-renewable-energy-target/.
26. Niti Aayog. 2018. "National Strategy for Artificial Intelligence–#AIforall." *Discussion Paper,* June 2018. https://niti.gov.in/writereaddata/files/Document publication/National Strategy.

27. Staeder, Tracy. 2017. "Modular Power Blocks Snap Together to Scale Up Energy Needs in Remote Areas." *IEEE Spectrum*, December 1, 2017.
28. Irena.org. 2019. "Innovation Landscape for a Renewable-Powered Future: Solution to Integrate Variable Renewables." www.irena.org/-/media/Files/IRENA /Agency/Publication /2019/Jan/IRENA_Innovation_Landscape_preview_2019.
29. Popescu, Adam. 2019. "AI Helps Africa Bypass the Grid." *Bloomberg.com*, June 11, 2018. www.bloomberg.com/news/articles/2018-06-11/aihelps-africa-bypass-the-grid.
30. Fickling, David. 2019. "Cyberattacks Make Smart Grids Look Pretty Dumb." *Bloomberg.com*, June 17, 2019. www.bloomberg.com/opinion/articles/2019–06–17/ argentina-blaming-hackers-for-outage-makes-smart-grids-look-dumb.
31. Mahendra, Ravi. 2019. www.smart-energy.com/industry-sectors/new-technology/ ai-is-the-new-electricity/.
32. Lydersen, K. and M. Kari. 2012. Smart Meters: Smart, But Not the Same as a Smart Grid." *Energy News Network*. https://energynews.us /2012/10/24/midwest/ smart-meters-smart-but-not-the-same-as-a-smart-grid.
33. Medium. 2019. "Azuri Technologies Receives $26 million Equity Investment to Expand its PAYG Solar Solutions Across Africa." June 26, 2019. https://medium.com/ techloy/azuri-technologies-receives-26-million-equity-investment-to-expand-its-payg-solar-solutions-bd48bd5c6226.
34. CBInsights. 2018. "5 Ways the Energy Industry is Using Artificial Intelligence." *Research Briefs*, March 8, 2018. www.cbinsights.com/research/artificial-intelligence-energy-industry.
35. Richter, Alexander. 2018. "Toshiba Energy Systems & Solutions Corporation Has Launched a Research Program on Internet-of-Things and Artificial Intelligence Technology to Improve the Efficiency of Geothermal Power Plants." *Think GeoEnergy*, August 16, 2018. www.thinkgeoenergy.com/.
36. Improving-efficiency-of-geothermal-plans-with-artificial-intelligence-and-iot-technolog.www.iitk.ac.in/npsc/Papers/NPSC1998/p47.pdf.
37. Vaughan, Adam. 2018. "AI and Drones Turn an Eye Towards UK's Energy Infrastructure." *theguardian.com*, December 2, 2018. www. theguardian.com/business/2018/ dec/02/ai-and-drones-turn-an-eye-towards-uks-energy-infrastructure.
38. Hodson, Hal. 2013. "AI Systems Switch Your Energy Bills to Save You Money." *Newscientist.com*, January 13, 2013. www.newscientist.com/article/mg21929315-200-ai-systems-switch-your-energy-bills-to-save-you-money/Gagan, Olivia. 2018.
39. Fitch, Asa. 2020. "The Key to Keeping the Lights On: Artificial Intelligence." *WSJ.com*, February 7, 2020. www.wsj.com/articles/the-key-tokeeping-the-lights-on-artificial-intelligence.
40. Warren, Chris. 2019. "Can Artificial Intelligence Transform the Power System?" *eprijournal.com*, January 29, 2019. http://eprijournal.com/can-artificialintelligence-transform-the-power-system/.
41. Christofaro, Beatrice. 2019. "Cyberattacks Are the Newest Frontier of War and Can Strike Harder Than a Natural Disaster. Here's Why the US Could Struggle to Cope If It Got Hit." *businessinsider.com*, May 23, 2019. www.businessinsider.com/ cyber-attack-us-struggle-taken-offline-powergrid-2019-4.

Chapter 8

Role of Artificial Intelligence and Machine Learning in Power Systems with Fault Detection and Diagnosis

Anjali Nighoskar, Shivani Gautam, and Kamini Lamba

Contents

DOI: 10.1201/9781003301820-8

8.1 Introduction

Energy is a critical component of global socioeconomic development. Global energy usage has increased in recent years. The limits imposed by large-scale fossil energy extraction and consumption endanger human evolution and existence. The economic advantages of knowledge automation are even more astounding. In 2013, McKinsey & Company's McKinsey Global Institute published research predicting that knowledge work automation would rank 2nd among the 12 disruptive technologies that would shape the future economy by 2025 [1]. Significant generating sources and loads pose new problems to traditional power system protection strategies. Intuitive and user-friendly prevention solutions based on enhanced measuring techniques and cognitive fault detection, such as ML, are beneficial in addressing these difficulties. Many research articles on ML-based power distribution problem diagnostics have been published. However, ML approaches are proliferating, and there is no comprehensive, up-to-date evaluation of ML-based power system problem diagnostics in the literature [2]. This chapter discusses the necessity for developing trends toward ML and a full assessment of ML-based power system failure diagnostics. First, an introduction to ML and AI is addressed, showing the various ML types and algorithms employed in Figure 8.1. Following that, Figure 8.2 shows that ML is a subset of AI. Figure 8.3 and Figure 8.4 illustrate the numerous uses of ML and AI in various fields. ML and AI are particularly beneficial in power systems and smart grid ways to tackle problems in these domains. Attempts were made to incorporate the flaws seen in traditional fault diagnosis, and resilience in power systems with different diagnostic techniques resulted in the rise of machine learning methodologies. A foundation structure and methodology for ML-based defect diagnostics are also described. Following that, many unsupervised and supervised learning algorithms used by researchers for fault detection and diagnosis are individually reviewed. Throughout the chapter, there are tabulated facts concerning approaches employed, the many simulation tools employed, and their advantages. Finally, the researcher's numerous concerns and challenges are elaborated with remedies, ideas, and future study chances.

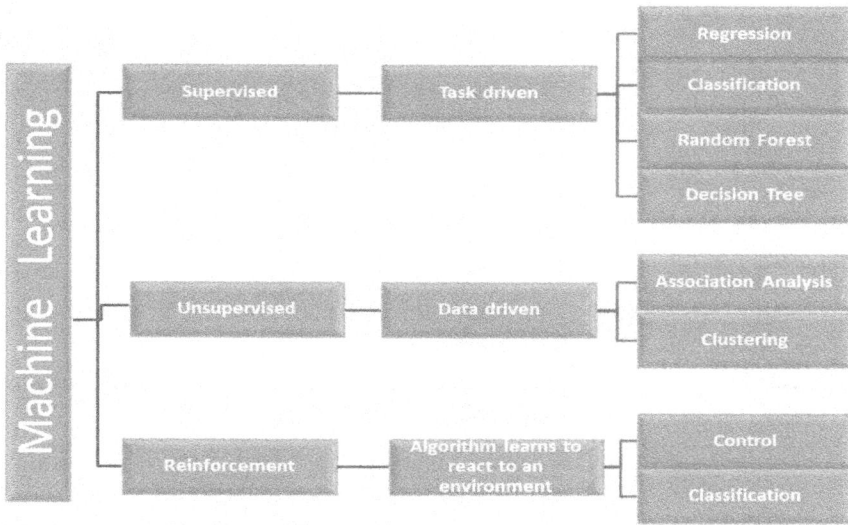

Figure 8.1   Machine Learning Types and Commonly Used Algorithms.

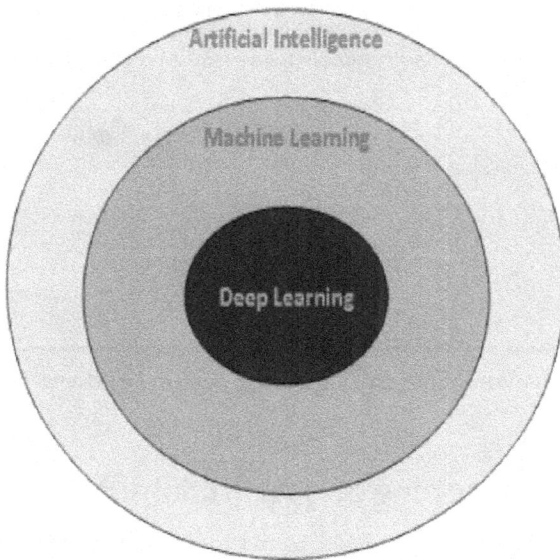

Figure 8.2   Machine Learning—Subset of Artificial Intelligence Techniques.

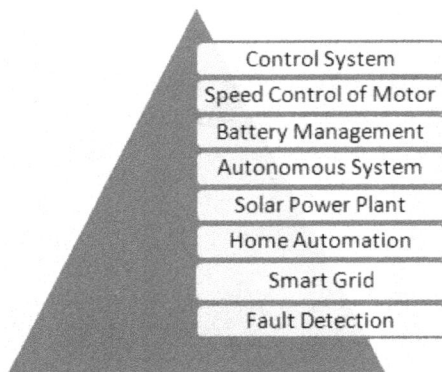

Figure 8.3   Applications of Machine Learning in Electrical Engineering.

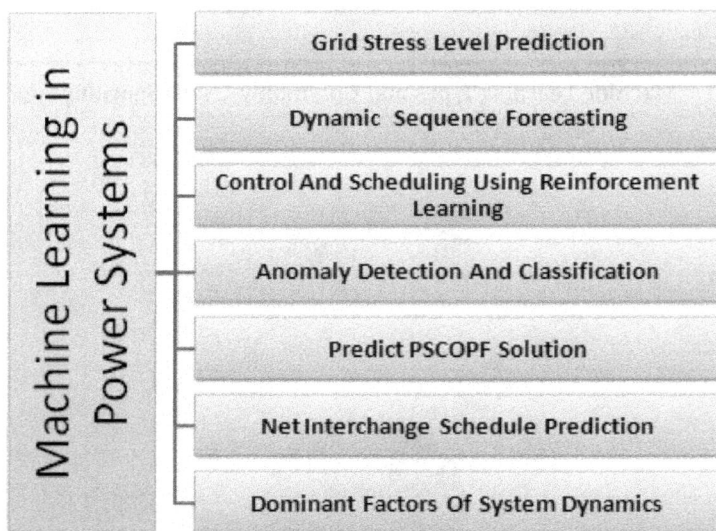

Figure 8.4   Application of Machine Learning in Power Systems.

8.2 Machine Learning and Artificial Intelligence

Machine learning is now widely used in the industry. Machine learning is being incorporated into more processes worldwide, and new possibilities for ambitious data scientists are multiplying. Based on the circumstances and nature of the floor, machine learning would assist the machine in determining the cleaning type, the cleaning intensity, and the cleaning duration [3]. In the last two decades, machine learning approaches have gotten much attention in domains including computer vision, information retrieval, and pattern recognition.

Machine learning and deep learning algorithms represent artificial intelligence approaches with various potential applications such as fault detection, voltage control and distribution analysis in power systems, character recognition, speech recognition, pattern identification, face recognition, multimedia data analysis, object detection, and power system research and applications. To increase the adaptability of machine learning approaches for power system applications, substantial work is necessary. With the fast advancement of technology, the popularity of Internet-based services, and the widespread usage of intelligent gadgets, many people are transmitting information through these devices. Artificial intelligence systems are critical in transmitting information from one location to another with accuracy and consistency across users [4].

Artificial intelligence approaches are a combination of machine learning and deep learning. These domains use intelligence models to anticipate and recognize patterns in data for autonomous reasoning [5]. The diagram (Figure 8.1) depicts a collection of artificial intelligence algorithms available through machine learning and deep learning approaches. AI 2.0 is the next generation of artificial intelligence (AI). Figure 8.2 shows that ML is the subset of AI.

8.3 Power System

A power system consists of a network of electronic components that provide, transmit, and utilize electricity. Among the numerous areas of electrical engineering, power system engineering has the most extended history of growth [6]. The electric power sector is undergoing severe, quick changes in both business and technological paradigms as it interacts with one of the largest customer-serving essential networks. As electricity systems get larger and more intricate as load demand grows, so does the demand for power plants, causing rising costs and rising generation of waste. End-to-end technologies such as industry 4.0, mechatronics components, wireless connectivity, artificial intelligence, and so on must be developed and widely used for energy digitization [7].

Electricity generation systems are composed of components that produce and distribute electricity to customers. A power generation and transmission system is made up of the following major components:

1. Power plants for generating electricity
2. Transformers to manage the r voltage levels as needed
3. Transmission lines for transmission of power
4. Power transmission substations and distribution lines
5. Transformers that reduce the voltage up to the required level by consumer electronics

The power is generated by converting the different available forms of energies (thermal, hydro, solar, wind, nuclear, fuels) into electrical energy.

Power systems in the modern day are controlled by highly competent personnel using automated controlled systems. A hierarchical control system's heart is the energy management system (EMS) that incorporates distant endpoint units, telecommunication networks, and multiple layers of high-performance computing. Numerous numerical optimization approaches have played an important role in cost reduction, operational dependability, and system security [8].

Machine learning (ML) applications have exploded in popularity in the field of power system research and development. Figure 8.3 depicts the application of ML in electrical engineering, and Figure 8.4 shows the application of ML in power system. Preliminarily, ML has now been implemented in smart grid (SG) and Energy Internet (EI), both of which are essential smart EEPS representations. Smart energy and electric power development systems will be accelerated by data-driven AI 2.0.

Electricity generation and transmission are very efficient and economical, but unlike other kinds of energy, electricity cannot be stored and must consequently be produced on demand.

8.4 Power System Fault

A power system fault is an exceptional situation that causes a large amount of current to flow through the equipment leading to disruption of a system's stability. Faults can be caused by a variety of factors, including climatic changes, mishandling, and mechanical failure [2, 3, 7]. The most common types of power system problems are classed as follows:

1. Open Circuit Fault
   Because the fault occurs in series with the transmission line, it is also known as a series fault. An open circuit defect is caused by the failure of one or two wires. It affects the reliability of the system. This sort of defect is further subdivided into Open Conductor, Two Conductor, and Three Conductor faults.
2. Short-Circuit Fault
   Excessive current flows in one or two system phases are a sort of defect because the electrodes of various phases communicate with one another via a transmission line, thyristor, or other circuit device. This type of faults can be further categorized as Symmetrical faults or balanced faults and Asymmetrical faults [4].

8.4.1 Causes, Effects, and Protection of Faults

The emergence of a power system fault can be caused by a variety of factors [5, 6, 7]. These could be caused by internal or external factors as described in Table 8.1.

Table 8.1 Different Types of Causes

Type of Causes	Explanation
Environmental conditions	Lightning, rain, snow, and other weather conditions like falling trees on the transmission line
Human errors	Inaccurate calculations, improper selection of electrical devices and equipment, delaying maintenance schedules
Failure of equipment	Insulation failures, aging of conductor, overloading of equipment
Other reasons	Vehicle collides with transmission line towers or poles, birds roosting on the lines, chemical pollution

Each form of breakdown has a different impact on the electrical system's operation. Some discrepancies are extreme and may result in a shutdown. Some of the effects of power system faults are summarized in Table 8.2.

There are tools that detect faults known as protection devices [2, 6]. These devices detect a malfunction and activate the circuit breaker, thereby disconnecting the defective circuit. The different protection devices include fuse, circuit breaker, protective relays, lightning arrestor, etc.

8.5 Smart Grid

The traditional power grid operations are restricted and inadequate. Because of the ineffective and inefficient power storage or reserve capacities, supply and demand must be balanced, and as a result, a forced just-in-time scenario is created. The smart grid is a solution to the transition to more sustainable technologies like distributed generation and microgrids. It aims to provide an effective method for combining renewable and green energy technology. The concept was proposed as a new perspective of the traditional infrastructure for the sake of meeting the growing need for energy. A smart grid, as defined by Industry 4.0, uses advanced future-oriented techniques (FoTs), intelligent information processing (IIP), and information and communication technologies (ICTs) to manage power generation, transmission, and distribution management in a much effectual, versatile, accurate, durable, distributed, protected, and cost-effective manner [9]. The typical centralized grid system may struggle to integrate and coordinate huge expanding interconnections. Figure 8.5 depicts the smart grid system's components.

Table 8.2   Different Types of Effects

Powerful fault current [2]	The fault produced a large magnitude fault current, which has the capacity to destroy the facilities in the power distribution system.
	Fault current causes overloading and mechanical stress in conductors.
	Strong current-induced arcing constitutes a serious risk of fire, which can damage other components if left unattended.
Overheating [2]	Insulation on conductors weakens and leads to reduced life.
Unbalanced current and voltage [2]	Causes heating of spinning equipment linked to the system.
	Generators may lose synchronism and cause the system to completely shut down.
	Affects the system stability.

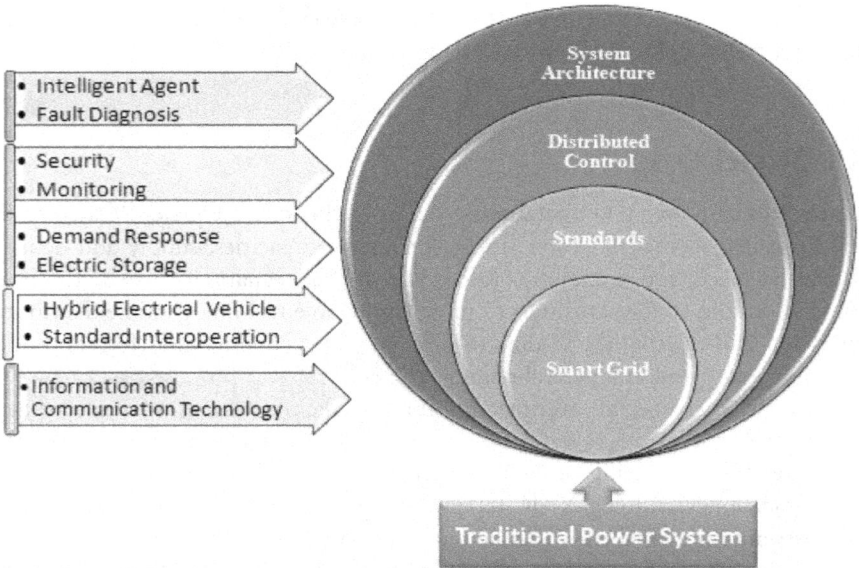

Figure 8.5   Smart Grid System.

The different advantages and applications of using the smart grid are listed as follows [10]:

1. Increases electric grid reliability by minimizing power quality disruptions and the repercussions and likelihood of widespread blackouts.
2. Allows for future technological breakthroughs and efficiencies.
3. Reduces consumer electricity prices by applying downward pressure.
4. Energy users continue to enjoy better affordability.
5. Consumers have more options for supply and knowledge.
6. Combines renewable and nonrenewable DERs.
7. Enhances security by lowering the impact and likelihood of natural disasters and man-made assaults.
8. Make alternate power generation sources more widely available.
9. Reduces the number of deaths and injuries caused by utility grid-related events, therefore reducing safety concerns.
10. Uses electric vehicles as both generators and storage devices, revolutionizing the transportation industry.
11. Increases total efficiency by decreasing energy losses and waste.
12. Smart grid decreases pollution by reducing greenhouse gas and carbon particle emissions, as well as providing cleaner power by encouraging the deployment of more renewable DERs.

8.6 Characteristics of Smart Grid

Smart grid incorporates cutting-edge products and services, as well as adaptive control, connectivity, surveillance.

1. The grid will be managed by a network of smart appliances, smart meters, micro-generation, energy storage, and consumer loads.
2. It permits different kinds and sizes of generators for connections and operations.
3. It supports various domestic micro-generation, storehouse solutions, and distributed generation.
4. It reduces the overall environmental effect of the electrical delivery system significantly.
5. It can deal with various cyber threats, physical attacks, and natural disasters, as well as deliver energy to customers with greater accuracy and protection in a self-healing manner.

Processing, alerting, self-healing, recovery plans, and security can all benefit from the Internet of Things (IoT). Integrating IoT with smart grid can benefit smart interfaces, detectors, meters, information-sharing devices, power plants (including biomass, coal, solar, and wind), energy backup, storage, consumption, and estimating necessary power to service clients. A number of IoT designs have been developed for smart grid integration.

They are classified as three-layer or four-layer designs [11–15]. Smart meters, network devices, and communication protocols are all part of Layer 1. Layer 2 devices are in charge of getting data from the central platform. Artificial intelligence systems are included in Layer 3 to supply information to billing systems. To utilize IoT in smart grid, we need to have certain technologies and meet certain conditions such as data fusion techniques, communication methods, reliability and ability of the IoT devices to work in the harsh environments, power transmission, and sensor types, etc.

8.7 Resilience of Power Systems

Techniques were employed to examine and improve the resilience of power systems that are grabbing attention. The word "resilience" refers to a power system's ability to "fight" and "restore" from a disruptive incident and also the ability to anticipate and respond to potential disruptive events and emerging threats [16]. Extreme weather occurrences (such as hurricanes, earthquakes, and flooding) are becoming more common. For example, since 1980, the United States has been devastated by 323 weather and climate disasters, with total reparation exceeding billion of US dollars (including CPI adjustment to 2022). The overall price of these 323 mishaps topped $2.195 trillion [17, 18].

Climate changes have caused widespread blackouts and significant damage to electricity infrastructure, resulting in losses to the economy and, more crucially, prolonged outage duration periods [19, 20].

Since the establishment of the resilience requirement, developing adequate optimization of power system resilience evaluation and upgrade methods has been a source of concern. Resilience of power systems can be viewed as an objective function, a constraint, or both. Methods generally such as linear regression, Mixed-Integer Linear Programming (MILP), Mixed-Integer Nonlinear Programming (MILNP), and Mixed-Integer Second-order Cone Programming (MISCOP), stochastic methods such as stochastic mixed-integer linear programming and stochastic mixed-integer nonlinear programming, and population-based intelligent search methods such as genetic algorithm (GA) are among the optimization methods [21]. These optimization problems are complex and difficult to compute. As a result, to simplify and successfully solve them, a proper decomposition approach is necessary. Common optimization solution methodologies include Column and Constraints Generation (C&CG), layered C&CG, Benders decomposition, greedy search algorithm, dual decomposition algorithm, scenario-based decomposition, and progressive hedging algorithm. These techniques have been tested in a number of different Integrated Development Environments (IDEs) such as GAMES and MATLAB® and solved by the different solvers such as IBM ILOG CPLEX Optimization Studio, Gaurobi, interior-point optimizer (IPOPT), and DDSIP.

8.8 Fault Detection and Diagnosis (FDD)

A fault is a deviation from a typical condition or behavior of a control input that is unsatisfactory. Unallowable deviation, in particular, denotes the distinctions between a given threshold and a fault value, which might result in a malfunction or failure. Faults are categorized on the basis of how they represent (additive or multiplicative defects) and their occurrence time (abrupt or stepwise faults, incipient or drifting faults, intermittent faults) [22]. A failure is defined as a continuous disturbance in a system's capability to meet necessary functions. The failure types may be categorized into three groups based on predictability: unpredictable (random), deterministic, and systematic or accidental failure. Finally, a breakdown or failure is defined as an unplanned interruption in the operation of a process or system [23].

The rising development of fault diagnostic systems has been prompted by the increasing demand to produce more dependable performance in complex systems like air navigation, computerized transportation, manufacturing, and power systems. The identification and diagnosis of faults is a difficult challenge in power system protection. It has recently been the focus of both business and academic research due to its importance in the dependability and safety of modern power systems. The major benefit of FDD is that it protects expensive repairs, shutdowns, putting human lives at risk, and equipment damage.

8.9 Classification Methods for Fault Detection and Diagnosis (FDD)

The FDD approaches can be utilized to construct energy systems, power transmission systems, industrial processes, machine components, and power transformers. All of these are composed of a number of controllers, actuators, sensors, and other devices. In the event of a problem, fault detection and classification approaches rely on changes in current and voltage signals. The FDD methods can be based on three different approaches as described in Figure 8.6. Methodologies for automatic problem identification and diagnosis in energy systems have advanced significantly, with remarkable advances utilizing data mining and machine learning (ML) approaches [47]. There are so many fault detection algorithms available for the fault analysis, the user may be unsure which one to employ. See Table 8.3. So, these methods are categorized as prominent techniques, hybrid techniques, and modern techniques.

Knowledge-driven and data-driven fault diagnostic methods are two important subcategories of fault diagnosis methods. Technologies in the knowledge-driven-based subcategory represent domain specialists' diagnostic reasoning. Techniques in the statistics segment, on the other hand, place a strong emphasis on pattern similarities [50].

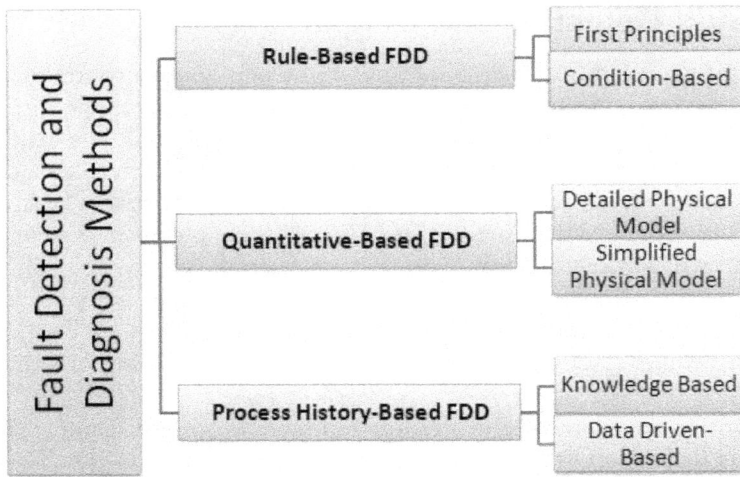

Figure 8.6   Fault Detection and Diagnosis Methods.

Table 8.3   Different Fault Classification Techniques

Common Techniques	• Artificial Neural Network (ANN) • Fuzzy Logic Approach • Wavelet Approach
Modern Techniques	• Functional Analysis and Computational Intelligence • Pattern Recognition Method • PCA-Based Method • Multi Information Measurement • Decision Tree-Based Method • GSM Method • Genetic Algorithm • SVM-Based Method
Hybrid Techniques	• Artificial Neural Network • Wavelet and Fuzzy Logic Technique • Fuzzy Logic Approach

8.10 Role of Machine Learning and ANN in FDD

To be safeguarded from transmission line problems, the modern power system needs real-time management and control. Early recognition of faults in a circuit may considerably improve system upkeep by preventing potentially dangerous damage from the problem. Recognizing the commencement of a micro fault in a power system automatically and precisely, particularly for failure severity degrees, remains a critical topic in the field of intelligent fault diagnostics. A pattern recognition approach might be beneficial in distinguishing between malfunctioning and healthy electrical power systems. ANNs are extremely effective in finding defective patterns and classifying faults using pattern recognition. Many algorithms based on ANN have been created, evaluated, and effectively applied in electrical power systems [24–29].

8.11 Comparison Based on Different Tools and Techniques

Table 8.4 shows the comparison based on different tools and techniques with the application area and characteristics.

8.12 Challenges and Opportunities of Power Systems

The issues in the realm of power systems can be classified as both technological and socioeconomic. The socioeconomic environment is critical to the deployment and success of any technology. If a technology fails to attract investors or consumers, it'll become irrelevant, resulting in the failure of major projects, the denial of advance ideas, and so on. Electricity load increase, energy crises, environmental pollution and climate change, unanticipated occurrences, aging infrastructures, and cyber problems are all significant concerns. Such challenges might occur as a result of economic or technological factors, as well as due to poor understanding among stakeholders [37, 38]. The challenges depend on the geographical regions, natural resources for generation, and the consumption of resources. A few most common challenges are listed here.

8.12.1 Deficiencies in the Grid's Framework

Electric grid infrastructure is still being built in developing countries such as India. The current grid network is inadequate to meet the predicted requirements for renewable power and distributed generation, which may offer a range of planning, building, implementation, and management challenges. It is heartening to learn that the government is addressing grid operational and connectivity challenges through its Central/State Transmission Utilities and National/Regional/State Load Dispatch Centre [39].

8.12.2 Electricity Storage Facility

There is a great need for power storage because the generation of electricity is not consistent and renewable sources are required for both bulk and distributed power

Table 8.4 Comparison Based on Different Tools and Techniques

Ref.	ML/AP Approach	Simulation Tool Used	Application Area	Characteristics
[6]	–	Mi Power Software (IEEE 14 bus system)	Power System	Evaluate the efficient way to analyze symmetrical faults
[26]	Artificial Neural Network	Synopsys CAD tools	Power Transmission Protection	They locate the faults and detected within one-third of a power cycle.
[30]	CNN and DWT	MATLAB®	Power System	Both algorithms find the fault detection in power system
[31]	Supervised back-propagation learning method and MLP	OpenDSS and MATLAB	Power System	Finds the fault detection faster and reliable
[32]	AI and Cyber Physical System (CPS)	MATLAB	Industry IoT System	They propose a robust intelligent control and prediction model
[33]	Auto Associative Neural Network (AANN) and Adaptive Neuro-Fuzzy Inference Systems (ANFIS)	Simulink®	Electrical Power System	It detects the faulty nodes with their type and the time of fault
[34]	Chaotic Neural Network and Wavelet Transform	MATLAB	Power System	Fault Detection of the Power System
[35]	Optimization-based consumption rescheduling model	Monte Carlo Simulink	Smart Power Grid	The vulnerability to attack is analyzed
[36]	Markov decision process-based detection algorithm	MATLAB and C programming	Smart Home power system	They reduced the bill of hackers and increased the peak energy usage in the local power system
[29]	ANN and Levenberg–Marquardt algorithm	MATLAB	Electrical power transmission system	They detect and classify the faults with good performance

generation in the smart grid systems. Different countries use different ways for storage; the US, Norway, Japan, India, and China, for example, use the pumped storage technique. The issue with pumped storage systems is that they necessitate enormous regions as reservoirs, which are often found only on mountain sides. In Germany, compressed air is used in subsurface storage, which can be used for generation of power [40]. Batteries are the most common kind of power storage, and lead-acid batteries are the most common type. Despite ongoing research to enhance efficiency and minimize the cost of battery technologies, battery storage options remain expensive.

8.12.3 Rising Electricity Demands

As the world's population has grown over the years, energy has become an increasingly important commodity. Most developed countries have a variety of energy-generating options to support various areas of the economy, but in developing countries like India and Africa, it is still a major problem to overcome the demand. A proportionate increase in energy output to fulfill the increasing demand is improbable due to the scarcity of resources [41].

8.12.4 Smart Grids and Communication Issues

Lots of effort have been made to transition the present electrical system into a smart grid, but smart grid companies will still confront three significant hurdles: ensuring standard interoperability, providing uninterrupted access to the unlicensed spectrum, and improving cyber security are among the issues. Because there are too many different technologies and standards on the market, the US Department of Energy, Office of Energy, Utilities and Natural Resources recommends identifying acceptable communication standards and developing interoperable communication protocols for smart grid [42]. Another significant problem is cyber security. While new applications for smart grid networks might increase efficiency and reliability, they can also introduce vulnerabilities if they are not implemented to meet security issues. Reliability and latency requirements, packet faults and changing connection capacity, severe climatic conditions, and resource limits are additional technological issues that smart grid faces while using Wireless Sensor Networks (WSNs) [48–49].

8.12.5 Cyber Security

In recent decades, there has been an increase in the number of cyber assaults targeting energy infrastructure, which have had a significant impact. Cyber assaults on smart grids can target the electricity transmission and distribution systems [43]. Cyber assaults on power distribution networks primarily aim to disrupt demand response by introducing fictitious electricity pricing [35–45]. Early identification and defense against cyber assaults are critical for reducing the effect of cyber attacks [36].

Recent research has used a variety of strategies to address the detection and protection of cyber threats. In power systems, real-time state estimation is used to identify incorrect data. Intelligent cyber assaults, on the other hand, may readily bypass the current state estimator, prompting the development of a slew of new methodologies, including machine learning approaches.

8.12.6 Energy Resources

Existing power stations mostly produce power using energy sources; petroleum, coal, nuclear fission, and natural gas are only a few examples. Because natural resources on the planet are nonrenewable, they are costly and depleted quickly [46]. The aging power grids that use fossil fuels emit greenhouse gas emissions and other chemicals that endanger both health and the environment. These recent growing challenges have prompted a worldwide search for long-lasting sustainable power resources, environmentally benign, and capable of supporting the power industry's long-term development for effective and high electricity.

8.13 Result and Discussion

AI 2.0 is currently at a tipping point, transitioning from "not practicable" to "can be practical," but there are still many barriers to overcome before the technology reaches the "extremely useful" level. ML, a typical algorithm classification in AI 2.0, generates projections and decisions by studying and gathering from huge volumes of combined archival and generated data and assists users in making accurate conclusions. ML technologies, advanced computer systems, and thorough data collection and analysis can be used in all areas of human activity. Today's computer systems recognize voices and features, unmanned vehicles, machine neural networks for trading activities, and so on because of the availability and growth of the digital revolution. Machine learning and deep learning algorithms represent artificial intelligence approaches with various potential applications such as fault detection, voltage control and distribution analysis in power systems, character recognition, speech recognition, pattern identification, face recognition, multimedia data analysis, object detection, and power system research and applications. Significant generating sources and loads pose new problems to traditional power system protection strategies. Artificial intelligence is giving significant assistance toward developing the intelligence revolution of the power and energy systems. Artificial intelligence combined with all other new technologies (IoT, ML, big data) can provide intelligent interactivity, safety, and controllability to the electrical system. The development and widespread usage of end-to-end technologies, such as the industrial Internet, robotics components, wireless communications, artificial intelligence, and so on, are required for the digitization of energy. ML has been used in the sectors of smart grid (SG) and Energy Internet (EI), both of which are significant smart EEPS representations. The electric power sector is undergoing severe, quick changes in both business and technological paradigms as it interacts with one of the largest customer-serving essential networks. AI 2.0, particularly ML, is entering a critical time of rapid development around the world and will play a critical role in smart EEPS. A vast number of research papers on ML-based power system problem diagnostics have been published. This chapter discussed the causes and types of power system faults. The different comparisons of reinforcement learning, deep learning, hybrid learning, etc. have been described in the SG and EI fields. Numerous unsupervised and supervised learning strategies that have been applied by several researchers for fault identification have been reviewed separately. The chapter also discussed the resilience with their optimization problem algorithm. The figures and tables show the different types of ML and also the tabulated data for fault detection, classification, and localization works with methodologies employed, various simulation tools used, and their application system. Attempts are made to identify the flaws existing in traditional fault detection, which resulted in the adoption of ML approaches. The chapter also described the progress and use of ML along with big data learning as well as the new prospects of AI 2.0 in smart energy and power systems. It examined the benefits and drawbacks of all fault diagnostic approaches, which will assist readers in selecting strategies for their study. Finally, the chapter identified research trends, major challenges, and future research areas.

8.14 Conclusion

The energy revolution will be accelerated by AI 2.0, ushering in a new era of smart energy. This chapter covers the entire power system, AI and machine learning methodologies, and various machine learning methods. It also provides an overview of various weather events and their effects on technology. This gives the researcher a better understanding of both the AI and ML technologies. The resilience of the power system gives the overview of different weather occurrences and their impact on the technologies. The different challenges and opportunities with different methods have been discussed. The tables give an overview of different technologies and tools used by the researchers. AI 2.0 and 3.0 technologies, particularly ML technologies, will be substantially integrated with SG and EI application strategies in the future. Smart sensor mixed with physical state, information learning associated with computer simulations, and secondary judgments connected with operational control are the three main combinations in the solution. As a consequence, from the viewpoints of theoretical approaches and innovation study, as well as application practices, the plans to collaborate boost the advancement of the next versions of AI, notably ML, and aid it to thrive in the smart sector.

References

1. R. P. Nath and V. N. Balaji. Artificial intelligence in power systems. *IOSR Journal of Computer Engineering (IOSR-JCE), e-ISSN* (2014): 2278–0661.
2. Loiy Rashed Almobasher and Ibrahim Omar A. Habiballah. Review of power system faults. *International Journal of Engineering Research & Technology (IJERT)* 9, no. 11 (November 2020).
3. Avagaddi Prasad, J. Belwin Edward and K. Ravi. A review on fault classification methodologies in power transmission systems: Part-I. *Journal of Electrical Systems and Information Technology*, Science Direct 5 (2018): 48–60.
4. Y. Y. Faruk Yalçın. A study of symmetrical and unsymmetrical short circuit fault analyses in power systems. *Sakarya University Journal of Science* 23, no. 5 (2019): 879–895.
5. A. S. Mubarak, A. S. Hassan, N. H. Umar, and M. Nasiru. An analytical study of power system under the fault conditions using different methods of fault analysis. *Advances in Electrical & Electronic Engineering* (2015): 113–119.
6. S. M. A. M. Singh. Analysis of symmetrical fault in IEEE 14 bus system for enhancing over current protection scheme. *International Journal of Future Generation Communication and Networking* 9, no. 4 (2016): 51–62.
7. S. S. R. K. R. P. N. A. X. Neha Kumari. Power system faults: A review. *International Journal of Engineering Research & Technology (IJERT)* 4, no. 02 (2016): 1–2.
8. Avagaddi Prasad, J. Belwin Edward and K. Ravi. A review on fault classification methodologies in power transmission systems: Part-II. *Journal of Electrical Systems and Information Technology* 5, no. 1 (2018): 61–67.
9. M. Faheem, S. B. H. Shah, R. A. Butt, B. Raza, M. Anwar, M. W. Ashraf, Md. A. Ngadi and V. C. Gungor. Smart grid communication and information technologies in the perspective of Industry 4.0: Opportunities and challenges. *Computer Science Review* 30 (2018): 1–30, ISSN 1574–0137.
10. G. Dileep. A survey on smart grid technologies and applications. *Renewable Energy* 146 (2020): 2589–2625, ISSN 0960–1481.
11. J. Lloret, J. Tomas, A. Canovas and L. Parra. An integrated IoT architecture for smart metering. *IEEE Communications Magazine* 54 (2016): 50–57.
12. C. Wang, X. Li, Y. Liu and H. Wang. The research on development direction and points in IoT in China power grid. Proceedings of the 2014 International Conference on Information Science, Electronics and Electrical Engineering, Sapporo City, Hokkaido, Japan, 26–28 April 2014; pp. 245–248.
13. X. Chen, L. Sun, H. Zhu, Y. Zhen and H. Chen. Application of internet of things in power-line monitoring. Proceedings of the 2012 International Conference on Cyber-Enabled Distributed Computing and Knowledge Discovery, Sanya, China, 10–12 October 2012; pp. 423–426.
14. S. K. Viswanath, C. Yuen, W. Tushar, W. T. Li, C. K. Wen, K. Hu, C. Chen and X. Liu. System design of the internet of things for residential smart grid. *IEEE Wireless Communications* 23 (2016): 90–98.
15. H. Ning and S. Hu. Technology classification, industry, and education for future internet of things. *International Journal of Communication Systems* 25 (2012): 1230–1241.
16. N. Bhusal, Michael Abdelmalak, Md Kamruzzaman and Mohammed Benidris. Power system resilience: Current practices, challenges, and future directions. *IEEE Access* 8 (2020): 18064–18086.
17. Smith, Adam B. (2019). NOAA National Centers for Environmental Information (NCEI) U.S. Billion-Dollar Weather and Climate Disasters. [Online]. Available: www.ncdc.noaa.gov/billions/.

18. Adam B. Smith and Richard W. Katz. US billion-dollar weather and climate disasters: Data sources, trends, accuracy and biases. *Natural Hazards* 67, no. 2 (2013): 387–410.
19. Crawford S. Holling. Resilience and stability of ecological systems. *Annual Review of Ecology and Systematics* 4, no. 1 (1973): 1–23.
20. Mathaios Panteli and Pierluigi Mancarella. Influence of extreme weather and climate change on the resilience of power systems: Impacts and possible mitigation strategies. *Electric Power Systems Research* 127 (2015): 259–270.
21. Yanlin Li, Kaigui Xie, Lingfeng Wang and Yingmeng Xiang. Exploiting network topology optimization and demand side management to improve bulk power system resilience under windstorms. *Electric Power Systems Research* 171 (2019): 127–140.
22. K. Severson, P. Chaiwatanodom and R. D. Braatz. Perspectives on process monitoring of industrial systems. *Annual Reviews in Control* 42 (2016): 190–200.
23. R. Isermann. *Fault-Diagnosis Systems: An Introduction from Fault Detection to Fault Tolerance*, 1st ed.; Springer: London, UK, (2006).
24. T. Dalstein and B. Kulicke. Neural network approach to fault classification for high-speed protective relaying. *IEEE Transactions on Power Delivery* 4 (1995): 1002–1009.
25. T. Bouthiba. Fault location in EHV transmission lines using artificial neural networks. *International Journal of Applied Mathematics and Computer Science* 14, no. 1 (2004): 69–78.
26. R. Venkatesan and B. Balamurugan. A real-time hardware fault detector using an artificial neural network for distance protection. *IEEE Transactions on Power Delivery* 16, no. 1 (2007): 75–82.
27. W. M. Lin, C. D. Yang, J. H. Lin and M. T. Tsay. A fault classification method by RBF neural network with OLS learning procedure. *IEEE Transactions on Power Delivery* 16, no. 4 (2001): 473–477.
28. M. S. Aziz, M. A. Abdel, M. Hassan and E. A. Zahab. High-impedance faults analysis in distribution networks using an adaptive neuro fuzzy inference system. *Electric Power Components and Systems* 40, no. 11 (2012): 1300–1318.
29. M. Jamil, S. K. Sharma and R. Singh. Fault detection and classification in electrical power transmission system using artificial neural network. *SpringerPlus* 4, no. 334 (2015).
30. Z. Wang and L. Xu. Fault detection of the power system based on the chaotic neural network and wavelet transform. *Complexity* 2020 (1 December 2020): 1–5.
31. Halil Alper Tokel et al. A new approach for machine learning-based fault detection and classification in power systems. 2018 IEEE Power & Energy Society Innovative Smart Grid Technologies Conference (ISGT). IEEE, 2018.
32. Zhihan Lv, Yang Han, Amit Kumar Singh, Gunasekaran Manogaran, and Haibin Lv. Trustworthiness in industrial IoT systems based on artificial intelligence. *IEEE Transactions on Industrial Informatics* 17, no. 2 (2020): 1496–1504.
33. Juan Pablo Nieto González, Luis E. Garza Castañion, Ruben Morales-Menendez, Ahmed El Hajjaji and Abdelhamid Rabhi. Fault diagnosis of electrical power systems using soft computing. *IFAC Proceedings Volumes* 46, no. 2 (2013): 863–868.
34. Zuoxun Wang and Liqiang Xu. Fault detection of the power system based on the chaotic neural network and wavelet transform. *Complexity* 2020 (2020).
35. Daogui Tang, Yi-Ping Fang, Enrico Zio and Jose Emmanuel Ramirez-Marquez. Resilience of smart power grids to false pricing attacks in the social network. *IEEE Access* 7 (2019): 80491–80505.
36. Yang Liu, Shiyan Hu and Tsung-Yi Ho. Leveraging strategic detection techniques for smart home pricing cyberattacks. *IEEE Transactions on Dependable and Secure Computing* 13, no. 2 (2015): 220–235.

37. Lefeng Cheng and Tao Yu. A new generation of AI: A review and perspective on machine learning technologies applied to smart energy and electric power systems. *International Journal of Energy Research* 43, no. 6 (2019): 1928–1973.

38. Lennyn Daza and Satyajayant Misra. Beyond the internet of things: Everything interconnected: Technology, communications and computing [book review]. *IEEE Wireless Communications* 24, no. 6 (2017): 10–11.

39. Ping Ju, Xiaoxin Zhou, Weijiang Chen, Y. Yu and C. Qin. "Smart grid plus" research overview. *Electric Power Automation Equipment* 38, no. 05 (2018): 2–11.

40. Jiye Wang, Kun Meng, Junwei Cao, Zhihua Cheng, Lingchao Gao and Chuang Lin. Information technology for energy internet: A survey. *Journal of Computer Research and Development* 52, no. 5 (2015): 1109.

41. J. Y. Wang, Y. Li, Z. M. Lu, W. X. Sheng and J. W. Cao. Research on local-area energy internet control technology based on energy switches and energy routers. *Proceedings of the Chinese Society of Electrical and Electronics Engineers* 36, no. 13 (2016): 3433–3439. https://doi.org/10.13334/j.0258-8013. pcsee.152857.

42. R. Ma, H. Chen, Y. Huang and W. Meng. Smart grid communication: Its challenges and opportunities. *IEEE Transactions on Smart Grid* 4, no. 1 (March 2013): 36–46. doi:10.1109/TSG.2012.2225851.

43. Muhammed Zekeriya Gunduz and Resul Das. Cyber-security on smart grid: Threats and potential solutions. *Computer Networks* 169 (2020): 107094.

44. D. Tang, Y. Fang, E. Zio and J. E. Ramirez-Marquez. Analysis of the vulnerability of smart grids to social network-based attacks 2018. 3rd International Conference on System Reliability and Safety (ICSRS) (IEEE), 2018; pp 130–134.

45. R. Tan, V. Badrinath Krishna, D. K. Yau and Z. Kalbarczyk. Impact of integrity attacks on real-time pricing in smart grids. Proceedings of the 2013 ACM SIGSAC conference on Computer & Communications Security, 2013; pp. 439–450.

46. Zhenya Liu. Research of global clean energy resource and power grid interconnection. *Proceedings of the CSEE* 36, no. 19 (2016): 5103–5110.

47. D. Baskar and P. Selvam. Machine learning framework for power system fault detection and classification. *International Journal of Scientific & Technology Research* 9 (2019).

48. Maria Lorena Tuballa and Michael Lochinvar Abundo. A review of the development of Smart Grid technologies. *Renewable and Sustainable Energy Reviews* 59 (2016): 710–725, ISSN 1364–0321.

49. Alireza Ghasempour. Internet of things in smart grid: Architecture, applications, services, key technologies, and challenges. *Inventions* 4, no. 1 (2019): 22. https://doi.org/10.3390/inventions4010022.

50. S. A. Aleem, N. Shahid and I. H. Naqvi. Methodologies in power systems fault detection and diagnosis. *Energy Systems* 6, no. 1 (2015): 85–108.

Chapter 9

AC Power Optimization Technique

Anita Soni and Anuprita Mishra

Contents

DOI: 10.1201/9781003301820-9

9.1 Introduction

Optimization utilizes important tools in solving problems of power distribution. As we are all aware, power production is less in comparison with demand. The optimization method can be applied to power system support to solve any power system operation, analysis, scheduling, and energy management issue and helps to solve the complex task of efficiently providing electricity to the grid [1, 2].

9.2 Power Sector

Artificial intelligence played a very important role in industrial development with power system expansion, stability, strengthening, and reliability. Technical advancements, selection, and dynamic response of the power system are essential. With the growth of the power system, complexity in. the networks is increased tremendously. Because of this power system analysis by conventional techniques, acquired data, process of information, management of remote devices, and utility [3] became more. complicated and time-consuming (Table 9.1).

AI is developed with the help of sophisticated computer tools and applied to resolve all problems described earlier for large power systems.

Fault prediction has been one of the major applications of artificial intelligence in the energy sector, along with real-time maintenance and identification of ideal maintenance schedules [4, 5]. In an industry in which equipment failure is common, with potentially significant consequences, AI combined with appropriate sensors can be useful to monitor equipment and detect failures before they happen.

Steady energy output is being discussed as a.potential source of base load power (the minimum amount of.power needed to be supplied to the electrical grid at any given time) to support the expansion [6–8] of less reliable renewable resources. Toshiba ESS has been conducting research on the use of IoT and AI to improve the efficiency and reliability of geothermal power plants. Predictive diagnostics enabled by rich data are used to predict problems that could shut down plants. Preventive measures such as chemical agent sprays to avoid turbine shutdowns are optimized (quantity, composition, and timing) using IoT and AI. Such innovations are

Table 9.1 Task Domains of Artificial Intelligence

Mundane (Ordinary) Tasks	Formal Tasks	Expert Tasks
Perception		
• Computer Vision	• Mathematics	• Engineering
• Speech Voice	• Geometry	• Fault Finding
	• Logic	• Manufacturing
	• Integration and Differentiation	• Monitoring
Natural Language Processing	Games	Scientific Analysis
• Understanding	• Go	
• Language Generation	• Chess (Deep Blue)	
• Language Translation	• Checkers	
Common Sense	Verification	Financial Analysis
Reasoning	Theorem Proving	Medical Diagnosis
Planning		Creativity
Robotics		
• Locomotive		

important in a country like Japan, which has the third-largest geothermal resources in the world, especially in the face of decreasing costs of competing renewable sources such as solar power.

The United Kingdom's National Grid used drones to monitor wires and pylons that transmit electricity from power stations to.homes and businesses [9–11]. Equipped with high-resolution still and infrared cameras, these drones. have been particularly useful in fault detection based on image.processing due to their ability to cover vast geographical areas and difficult terrain. They have been used to cover 7,200 miles of overhead lines. AI is then used to monitor the conditions of.power assets and determine when they need to be replaced or repaired. The digital transformation of home energy management and consumer appliances will allow automatic meters to use AI to optimize energy consumption and storage [12].

AI empowers consumers by allowing them to determine energy source, their household budget, or their consumption patterns. Researchers at Carnegie Mellon University have developed a [13, 14] machine learning system called Lumator that combines the customer's preferences and consumption data with information on the different tariff plans, limited-time promotional rates, and other product offers in order to provide recommendations for the most suitable electricity supply deal [15]. Such solutions can also help increase the share of renewable energy by helping consumers convert their preference for renewable energy into realized demand for it. They can also be used to signal to producers the level of consumer demand for renewable energy. AI will facilitate decision-making about optimal times for distributed generation to contribute to the grid, rather than draw from it [16]. AI can also assist traditional producers and system operators who will now have to balance increased intermittent renewables and distributed generation.

AI can be used to spot discrepancies in usage patterns, payment history, and other consumer data. Furthermore, when combined with automated meters, it can improve monitoring [17, 18]. It can also help optimize costly and time-consuming physical inspections. These informal connections and billing errors have benefited consumers and provided solutions by applying AI. Furthermore, the University of Luxembourg has developed an algorithm that analyzes information, from electricity meters to detect abnormal usage [19]. The algorithm managed to reveal problem cases at a higher rate than most other tools, when applied to information over five years from 3.6 million Brazilian households [20, 21]. The technology is slated to be deployed across Latin America. The main types of voltage problems are as follows:

1. Planning of system reactive power demand
2. Installation of reactive power control resources
3. The operation of existing voltage resources and control device

9.3 Agent and Environment

An **agent** is anything that can perceive its environment through **sensors** and acts upon that environment through **effectors**. See Figure 9.1.

- A **robotic agent** replaces cameras and infrared range finders for the sensors, and various motors and actuators for effectors.
- A **software agent** has encoded bit strings as its programs and actions [22, 23].

9.3.1 Agent Terminology

- **Performance Measure of Agent**: It is the criteria that determine an agent's success.
- **Behavior of Agent**: It is the action that an agent performs after any given sequence of precepts [24].
- **Percept**: It is an agent's perceptual inputs at a given instance.
- **Agent Function**: It is a map from the percept sequence to an action.

9.3.2 Rationality

Rationality is concerned with the expected status of being reasonable, sensible, and having good sense of judgment actions. and results depending on what the agent has perceived [26]. Performing actions. with the aim of obtaining useful information is an important part of rationality.

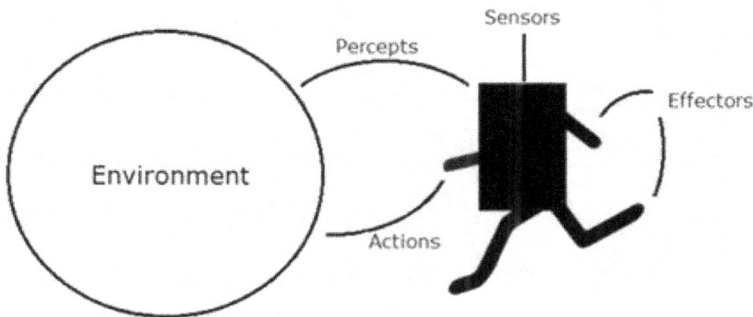

Figure 9.1   Environment Effect.

9.3.3 Ideal Rational Agent

The ideal rational agent expects actions to maximize its performance measure, based on the following:

- Its percept sequences.
- Its built-in knowledge bases.

The rationality of an agent depends on the following:

- The **performance measures**, which determine the degree of success
- The agent's **percept sequence** till now.
- The agent's **prior knowledge about the environment**.
- The **actions** that the agent can carry out.

A rational agent always performs the right action, where the right action means the action that causes the agent to be most successful in the given percept sequence [27]. The problem the agent solves is characterized by performance measure, environment, actuators, and sensors (PEAS).

9.3.4 Structure of Intelligent Agents.

- Agent = architecture + agent program.
- Architecture = the machinery that executes like an agent
- Agent Program = helps us in implementation of an agent function.

9.3.5 Simple Reflex Agents

- They choose actions based on the current percept.
- They are rational if a correct decision is made only based on current percept.
- Their environment is completely observable [28].

Condition-Action Rule is a rule that maps a state (condition) to an action. See Figure 9.2.

9.3.6 Model-Based Reflex Agents

We used model based on actions. They maintain an internal state. See Figure 9.3.

Model: It is knowledge about "how the things happen in the world."

Internal State: It is a representation of unobserved aspects of current state depending on percept history [29].

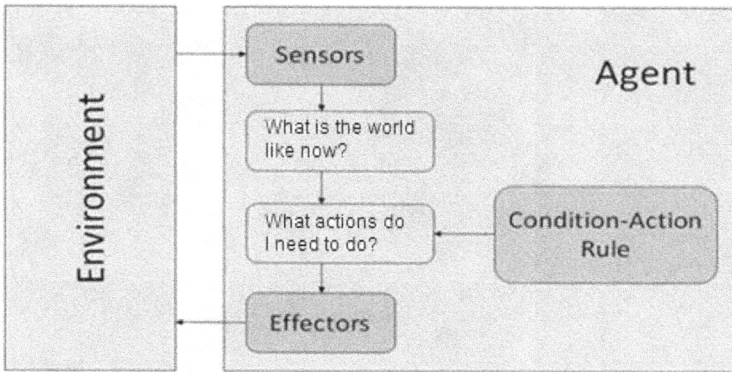

Figure 9.2 Environment Action.

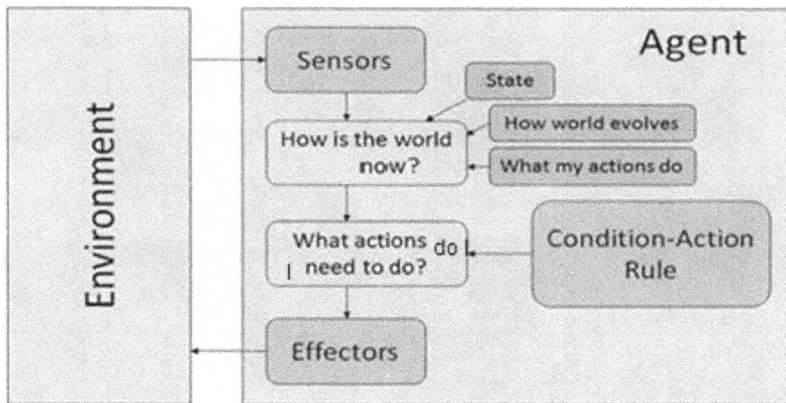

Figure 9.3 Detecting Agent Action in Environment.

Updating the state requires the information about the following:

- How the world evolves.
- How the agent's actions affect the world.

9.3.7 Goal-Based Agents

Goal: It is the description of desirable situations. Agents choose actions to achieve goals. The goal-based approach is more flexible than model based approach since

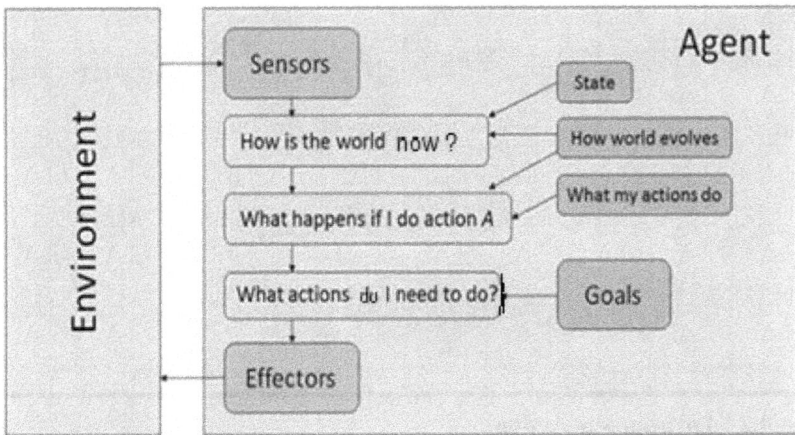

Figure 9.4   Detection of Agent Based on Goals.

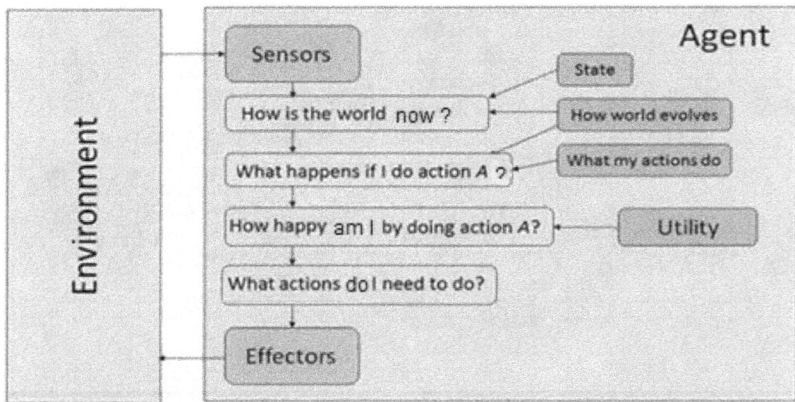

Figure 9.5   Selection of Agent Based on Utility.

its knowledge supporting a decision is explicitly [30] modeled, thereby allowing for modifications. See Figure 9.4.

9.3.8 Utility-Based Agents

The agents choose actions based on a preference (utility).for each state. See Figure 9.5.
Goals are inadequate when –

■ There are conflicting goals, out of which only few can be achieved.
■ Goals have some uncertainty of being achieved and you need to weigh the likelihood of success against the importance of a goal [31].

9.4 Why Fuzzy Logic?

Fuzzy logic is useful for commercial and practical purposes.

- It can control machines and consumer products.
- It may not give accurate reasoning, but acceptable reasoning.
- Fuzzy logic helps to deal with uncertainty in engineering.

9.4.1 Fuzzy Logic Systems Architecture

It has four main parts as shown next:

- **Fuzzification Module**: It transforms the system inputs, which are crisp numbers, into fuzzy sets. It splits the input signal [32] into five steps (see Table 9.2):
- **Knowledge Base**: It stores *if-then* rules provided by experts.
- **Inference Engine**: It simulates the human reasoning process by making fuzzy inference on the inputs and *if-then* rules.
- **Defuzzification Module**: It transforms the fuzzy set obtained.by the inference engine into a crisp value [33]. See Figure 9.6.

For reactive power control with the objective of enhancing the voltage profile of power system, fuzzy logic has been applied. The voltage deviation and controlling variables are converted into fuzzy set, or fuzzy system notations to construct the relations between voltage deviation and controlling ability of the controlling device [34, 35]. The main control variables are generator, excitation, transformer taps, and VAR compensators. A fuzzy system is formed to select these control variables and their movement [36, 37].

The control variables are selected based on the following:

1. Local controllability toward a bus having unacceptable voltage
2. Overall controllability toward the buses having poor voltage profile 2.4 [38]

Table 9.2   Input Signal Value

Inputs	Inputs Signal
LP	X is Large Positive
MP	X is Medium Positive
S	X is Small
MN	X is Medium Negative
LN	X is Large Negative

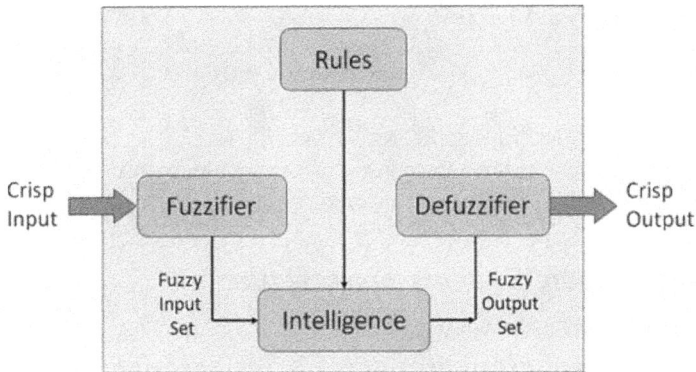

Figure 9.6 I/O Fuzzification.

How fuzzy logic can be used in power systems:

Fuzzy logic can be used for designing the physical components of power systems. As most of the data used in power, system analysis is approximate values and assumptions; fuzzy logic can be of great use to derive a stable, exact, and ambiguity-free output [39].

Artificial neural networks and expert systems can be used to improve the performance of power supply in line. The environmental sensors sense the environmental and atmospheric conditions and give them as input to the expert systems. The expert systems are computer programs written by knowledge engineers, which provide the value of line parameters to.be deployed [40]. The ANNs are trained to change the values of line parameters over the given ranges based on the environmental conditions. After training is over, neural network is tested and the performance of updated trained neural network is evaluated. If performance is not up to the desired level, some variations can be done like varying the number of hidden layers and varying the number of neurons in each layer. The processing speed is directly proportional to the number of neurons. These networks take different neurons for different layers and different activation functions [41, 42].

9.4.2 Looking to the Future

AI helps the energy sector in both emerging markets and advanced economies, which continue to face multiple challenges in terms of efficiency, transparency, affordability, and the integration of, renewable energy sources in power systems [43].

First, AI companies have expertise in math and computer science, but they often lack the knowledge needed to understand the specifics of power systems. And this problem is more acute in emerging markets. While the potential applications of

AI in the power sector are multiple and varied, there is a need to educate the AI industry more deeply on the aspects of the power sector. Cloud-based applications are widespread and central to AI solutions, but there are regulatory restrictions on their use in the power industry. This is changing, however, as the benefits of AI cloud applications become more evident [44].

Second, AI's potential is not perfect in rural and other underserved areas in many emerging markets, particularly low-income countries. Smart meters rely on constant data communication, so a lack of reliable connectivity is a substantial impediment in areas where cellular network coverage is sparse or limited [45].

Third, digital transformation of the power grid helps in detecting cyber attacks on critical infrastructure, which can be as damaging as a natural disaster. The growing threat from hacking has become common and a matter of significant concern, particularly because smart metering and automated control have come to represent close to 10% of global grid investments, equivalent to $30 billion a year dedicated to digital infrastructure [46–48].

Fourth, data sources and ensuring representativeness given the diversity within the data will be challenging. Other challenges may also arise because of a low volume of data for machine learning models to learn from. Contextualization and transfer of learning of two similar tasks could prove to be difficult. Furthermore, these models could be susceptible to inaccurate data. These challenges are being partly addressed through reinforcement learning [49].

Fifth, AI-based models are essentially black boxes to their users. Since existing models are far from perfect, it is necessary to have safe guards in place when incorporating, them into energy systems. When combined with better analytics, sensors, robotics, and IoT, devices, AI can be used for automation of simple tasks, allowing humans to focus on the unstructured challenges [50].

Sixth, Imbalance in priorities and therefore, investments in smart meters is more profitable than smart grids, smart meters are decision-making tools for customer choice. Customers can decide when to turn their power on or off, or change their consumption habits during peak times. Smart grids, by contrast, are less about the consumer and more about making quick adjustments to ensure the electricity flows as efficiently as possible [51, 52].

9.4.3 Practical Implementation of AI Method in AC

Fans blow air over your skin, which promotes moisture evaporation. Evaporation is an endothermic process, meaning the moisture absorbs heat from our surroundings as it turns into vapor. Air conditioners (ACs) utilize the same mechanism but in a far more complex way. Inside your air conditioner is a coil of coolant that's continuously undergoing evaporation and condensation. Air is pulled into the air conditioner and cooled by the evaporation process [53, 54]. In this process, condensation radiates heat outside your home. See Figure 9.7.

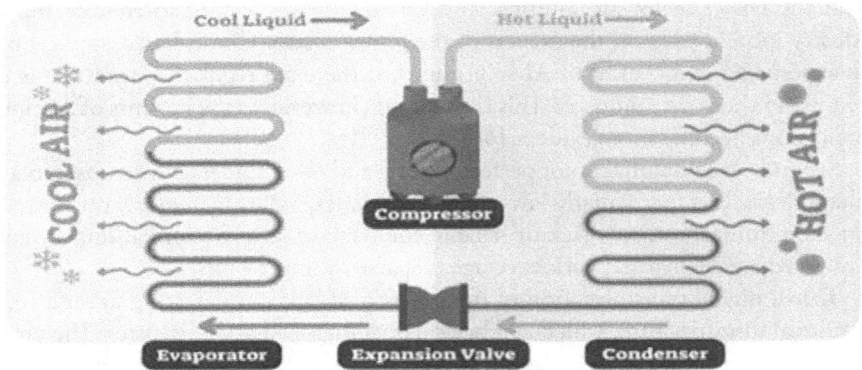

Figure 9.7 Air Conditioner (AC) Working.

9.4.4 Factors Affecting AC Power Consumption

1. **Number of people in the room**: The human body emits lot of heat; a room with 20 people will need a bigger size AC or multiple ACs to cool the air inside it, whereas the same room with three people in it will get the same cooling for a smaller size AC or a single AC [55]. So, more people mean more power; hence, a higher electricity bill. See Figure 9.8.
2. **Outside and inside temperature**: It takes more power to cool a room when outside temperature is 40 degrees Celsius than when it is 32 degrees Celsius. Similarly, it takes more power to cool at room to 18 degrees than to cool it at 24 degree Celsius.
3. **Room size**: Electricity consumed in removing heat from a 100 sqft room is less than removing heat from a 200 sqft room. Air conditioners remove the heat from the air, inside your room. Therefore, the larger the room, the larger is the volume of air inside it and more electricity is required to cool that air [55].
4. **Electrical appliances in your room**: Every electrical appliance in your room generates, heat, which increases your room's temperature and makes your AC do additional work for cooling the same volume of air, hence more power consumption (Table 9.3).
5. **Volume of air to be cooled**: Oftentimes, we keep our cupboard doors open, which increases the volume of air to be cooled by your AC [56]. Now your AC has to cool the hot air inside your cupboard too, which is unnecessary and consumes extra power.
6. **Objects in your room**: Every solid object in your room gets cold when kept in AC; this is another unnecessary cooling your AC has to do.

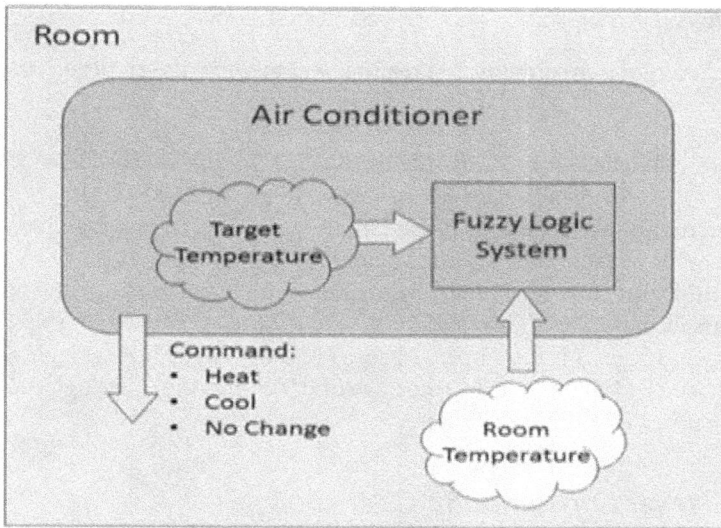

Figure 9.8 Air Conditioner (AC) Effects in Room Temperature.

Table 9.3 Matrix of Room Temperature versus Target Temperature

Room Temp/ Target	Very Cold	Cold	Warm	Hot	Very Hot
Very Cold	No-Change	Heat	Heat	Heat	Heat
Cold	Cool	No-Change	Heat	Heat	Heat
Warm	Cool	Cool	No-Change	Heat	Heat
Hot	Cool	Cool	Cool	No-Change	Heat
Very Hot	Cool	Cool	Cool	Cool	No-Change

Calculate Power Consumption of an AC Manually

1. **The wattage of AC:** The unit of power used by AC, that is, the consumption rate of device.
2. **Operational hours:** The total number of hours AC running.
3. **Electricity tariff:** The rate of 1 unit (kWh) of electricity.

9.4.5 Key Points

The calculation of a circuit's wattage requires you to know the amperage and voltage of the circuit.

- Determine the voltage of the circuit. Assume, for this example, that the circuit has 110 volts, the standard house voltage in the United States.
- Determine the amperage in the circuit. Assume, for this example, that the circuit has a light bulb drawing about 0.91 amps.
- Multiply the number of amps by the number of volts to obtain the number of watts in the circuit. The equation W = A x V shows this relationship where W is the wattage, A is the amperage, and V is the voltage. This example assumes a voltage of 110 volts and an amperage of 0.91 amps [56]. The light bulb therefore uses 110 x 0.91 = 100 watts.

9.4.5.1 Power Consumption Calculator

Power consumption of 1.5 TR Split AC—1500 Watts. You can check this power consumption of your AC model on the BEE sticker pasted on the indoor unit.

- Let consider AC running = 8 Hrs Daily
- Total units consumed = 1500 Watts or 1.5 kw multiply by 8 Hrs
  = 1.5 Kw * 8 Hrs = 12 kwh

This is the total amount of units of electricity (AC is going to consume), but the compressor of the AC is not going to run 8 hrs daily, it's going to shut down after the desired temperature is approached.

- Considering compressor is going to run 70% only then total units become
  = 70% *12 kwh = 8.4 Kwh
  Total units consumed = 8.4 KWH
- Now you can multiply this factor by the price per unit in your area.
- Let us suppose cost = Rs 8/Unit
  = Rs 8 * 8.4 kwh = Rs 67.2/- (running AC 8 Hrs Daily)

Outside temperature will affect consumption of AC when the door is open.

So, the compressor is not going to cut off and will run on full speed throughout. You can take 12 kwh units' consumption in that case [57].

9.4.6 Practical Approach AI for Minimizing AC Power Consumption

Step 1: Define Linguistic Variables and Terms
Maintain room temperature;, cold, warm, hot, etc., are linguistic terms.

Temperature (t) = {very-cold, cold, warm, very-warm, hot}

Every member of this set is a linguistic term, and it can cover some portion of overall temperature values (Table 9.4).

Step 2: Construct Membership Functions for Them

The membership functions of temperature variables are as shown in Figure 9.9.

Step 3: Construct Knowledge-Based Rules

Step 4: Obtain Fuzzy Value

Fuzzy set operations perform evaluation of rules. The operations used for OR and AND are Max and Min, respectively. Combine all results of evaluation to form a final result. This result is a fuzzy value.

Step 5: Perform Defuzzification

Defuzzification is then performed according to membership function for output variable. By using this approach, we are able to evaluate accurate consumption of electricity based on temperature and function, so we can optimize our consumption. See Figure 9.10.

Table 9.4   Structure of If-Then-Else

S. No	Condition	Action
1	If temperature = (Cold OR Very Cold) AND target = Warm Then	Heat
2	If temperature = (Hot OR Very Hot) AND target = Cool Then	Cool
3	If (temperature = Warm) AND (target = Warm) Then	No-Change

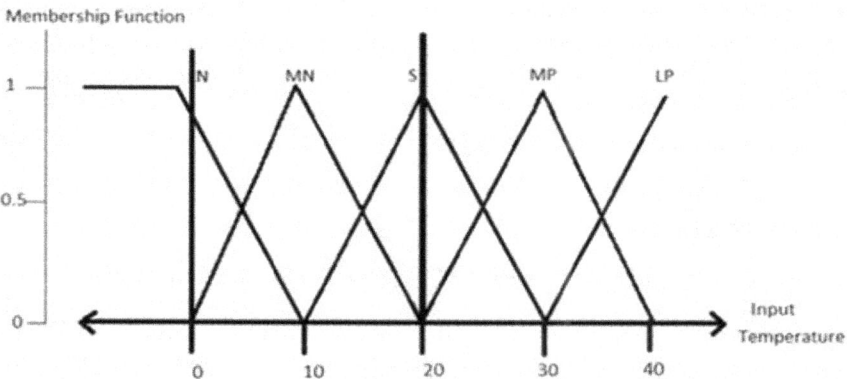

Figure 9.9   Effect of Input Signals with Functions.

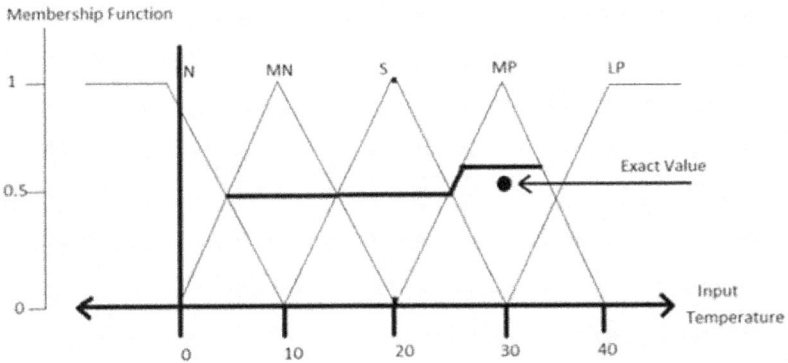

Figure 9.10 Evaluating Exact Value Based on Temperature and Function.

9.5 Summary

Every optimization problem has three components: an objective function, decision variables, and constraints. When one talks about formulating an optimization problem, it means translating a "real-world" problem into the mathematical equations and variables that comprise these three components. The objective function, often denoted f or z, reflects a single quantity to be either maximized or minimized. Examples in the transportation world include "minimize congestion," "maximize safety," "maximize accessibility," "minimize cost," "maximize pavement quality," "minimize emissions," "maximize revenue," and so forth. You may object to the use of a single objective function.

We are not able to produce power, it's very complex and that's why it's necessary to save power. Air conditioners operate in a continuous cycle that involves compression, condensation, expansion, and evaporation. Outside the home, the air conditioner compresses the gaseous refrigerant, which increases its temperature. A fan blows outside air across the unit's coils containing the hot, high-pressure refrigerant. When the outside air is more relaxed than the fluid, heat energy flows from the refrigerant to the outside air. When the high-temperature gaseous refrigerant gives up energy, it turns back into a liquid.

References

1. Kwang Lee and H. Elsharkawi. (2008). *Modern Heuristic Optimization Techniques.* Wiley.
2. Debirupa Hore and N. M. Lokhande. (2013). Computational Analysis of Different Artificial Intelligence Based Optimization Techniques for Optimal Power Flow and Economic Load Dispatch Problems. *International Journal of Computers & Technology,* vol. 4, January–February.

3. Keyan Liu, Wanxing Sheng, and Yunhua Li. (2006). Research on Reactive Power Optimization Based on Adaptive Genetic Simulated Annealing Algorithm. International Conference on Power System Technology, Chongqing, pp. 1–6.
4. Jingui Lu, Li Zhang, and Hong Yang. (2010). Combining Strategy of Genetic Algorithm and Particle Swarm Algorithm for Reactive Power Optimization. International Conference on Electrical and Control Engineering (ICECE-2010), Wuhan, pp. 3613–3616.
5. Altaf Badar, Dr. B. S. Umre, and Dr. A. S. Junghare. (2014). Study of Artificial Intelligence Optimization Techniques Applied to Active Power Loss Minimization. *Journal of Electrical and Electronics Engineering*, pp. 39–45.
6. S. Binitha and S. Siva Sathya. (2012). A Survey of Bioinspired Optimization Algorithm. *International Journal of Soft Computing and Engineering (IJSCE)*, vol. 2.
7. O. P. Malik. (2004). Amalgamation of Adaptive Control and AI Techniques: Applications to Generator Excitation Control. *Annual Reviews in Control*, vol. 28, pp. 97–106.
8. G. Hwang, D. Kim, J. Lee, and Y. An. (2008). Design of Fuzzy Power System Stabilizer Using Adaptive Evolutionary Algorithm. *Engineering Applications of Artificial Intelligence*, vol. 21, pp. 86–96.
9. V. Mukherjee and S. P. Ghoshal. (2007). Intelligent Particle Swarm Optimized Fuzzy Pid Controller for AVR System. *Electrical Power Systems Research*, vol. 77, no. 12, pp. 689–1698.
10. M. Caner, N. Umurkan, S. Tokat, and S. V. Ustun. (2008). Determination of Optimal Hierarchical Fuzzy Controller Parameters According to Loading Condition with ANN. *Expert Systems with Applications*, vol. 34, pp. 2650–2655.
11. SP Ghoshal, A. Chatterjee, and V. Mukherjee. (2009). Bio-Inspired Fuzzy Logic Based Tuning of Power System Stabilizer. *Expert Systems with Applications*, vol. 36, pp. 9281–9292.
12. Altaf Badar, Dr. B. S. Umre, and Dr. A. S. Junghare. (2014). Study of Artificial Intelligence Optimization Techniques Applied to Active Power Loss Minimization. *Journal of Electrical and Electronics Engineering*. e-ISSN: 2278-1676, p-ISSN: 2320-3331.
13. T. Yuvaraja, K. Ramya, and M. Gopinath. (2015). Artificial Intelligence and Particle Swarm Optimization Algorithm for Optimization Problem in Microgrids. *Asian Journal of Pharmaceutical and Clinical Research*, vol. 8, no. 3, January.
14. Sina Khajeh Ahmad Attari, Mahmoud Reza Shakarami, and Farhad Namdari. (2016). Pareto Optimal Reconfiguration of Power Distribution Systems with Load Uncertainty and Recloser Placement Simultaneously Using a Genetic Algorithm Based on NSGA-II. *Indonesian Journal of Electrical Engineering and Computer Science*, vol. 1, March, pp. 419–430.
15. S. Harish Kiran, Subhransu Sekhar Dash, and C. Subramani. (2016). Performance of Two Modified Optimization Techniques for Power System Voltage Stability Problems. *Alexandria Engineering Journal*, vol. 55, no. 3.
16. Amr K. Khamees, Ahmed El-Rafei, N. M. Badra, and Almoataz Y. Abdelaziz. (2017). Solution of Optimal Power Flow Using Evolutionary-Based Algorithms. *International Journal of Engineering, Science and Technology*, vol. 9, pp. 55–68.
17. Chandragupta Mauryan Kuppamuthu Sivalingan, Subramanian Ramachandran, and Purrnimaa Shiva Sakthi Rajamani. (2017). Reactive Power Optimization in a Power System Network Through Metaheuristic Algorithms. *Turkish Journal of Electrical Engineering & Computer Sciences*, vol. 25, no. 6, Article 14. https://doi.org/10.3906/elk-1703-159.

18. Ö. P. Akkaş, E. Çam, İ. Eke, and Y. Arıkan. (2018). New Optimization Algorithms for Application to Environmental Economic Load Dispatch in Power Systems. *Electrica*, vol. 18, no. 2, pp. 133–142.
19. Sarah Dean, Horia Mania, Nikolai Matni, Benjamin Recht, and Stephen Tu. (2018). *On the Sample Complexity of the Linear Quadratic Regulator.* California Institute of Technology, December 17.
20. C. T. M. Clack, Y. Xie, and A. E. MacDonald. (2015). *Linear Programming Techniques for Developing an Optimal Electrical System Including High-Voltage Direct-Current Transmission and Storage.* Electrical Power and Energy Systems.
21. P. K. Dash, A. C. Liew, and S. Rahman. (1996). Fuzzy Neural Network and Fuzzy Expert Systems for Load Forecasting. *IEE Proceedings—Generation Transmission and Distribution*, vol. 143, no. 1, pp. 106–114.
22. K. H. Kim, K. J. Park, K. J. Hwang, and S. H. Kim. (2005). Implementation of Hybrid Short Term Load Forecasting Using Artificial Neural Networks. *International Journal of Emerging Electric Power Systems*, vol. 2. DOI: 10.2202/1553–779X.1021.
23. Y. M. Park, J. B. Park, and J. R. Won. (1997). A Hybrid Genetic Algorithm/Dynamic Programming Approach to Optimal Long-Term Generation Expansion Planning. *International Journal of Electrical Power and Energy Systems*, vol. 20, no. 4, pp. 295–330.
24. A. Afzalin and D. A. (2000). Linkens. Training of Neuro Fuzzy Power System Stabilizer using Genetic Algorithm. *International Journal of Electrical Power and Energy Systems*, vol. 22, no. 2, pp. 93–102.
25. W. S. Jwo, C. W. Liu, C. C. Liu, and Y. T. Hsiao. (1995). Hybrid Expert System and Simulated Annealing Approach to Optimal Reactive Power Planning. *IEE Proceedings, Generation Transmission and Distribution*, vol. 142, no. 4, pp. 381–385.
26. F. Benhamida, Y. Salhi, I. Ziane, S. Souag, R. Belhachem, and A. Bendaoud. (2013). A PSO Algorithm for The Economic Load Dispatch Including a Renewable Wind Energy. 3rd International Conference on Systems and Control, pp. 1104–1109.
27. K. P. Wong and S. Y. W. Wong. (1997). Hybrid Genetic/Simulated Annealing to Short Term Multiple Fuel-Constrained Generation Scheduling. *IEEE Transactions Power Systems*, vol. 12, no. 2, pp. 776–784.
28. H. Kim, K. Nara, and M. Gen. (2004). A Method for Maintenance Scheduling Using GA Combined with SA. *Computers and Industrial Engineering*, vol. 27, no. 4, pp. 477–480.
29. K. Wardwick, A. Ekwue, and R. Aggarwal. (1997). *Artificial Intelligence Techniques in Power Systems.* IEE.
30. Dan Cristian, Constantin Barbulescu, Stefan Kilyeni, and Vasile Popescu. (2013). Particle Swarm Optimization Techniques-Power Systems Applications. Proceedings of 6th IEEE International Conference on Human System Interaction (HIS).
31. R. Rekha and G. Kannan. (2013). A Comparative Analysis on Reactive Power Optimization Using Various Techniques in Deregulated Power System. Proceedings of IEEE 2013, International Conference on Energy Efficient Technologies for Sustainability (ICEETS).
32. S. A. Jumaat, I. Musirin, M. M. Othman, and H. Mokhlis. (2011). A Hybrid Meta Heuristics Optimization Technique for Loss Minimization and Cluster Identification in Power System Network. Proceedings of 2011, International Conference on Energy Efficient Technologies for Sustainability (ICEETS).
33. Himmat Singh, Laxmi Srivastava. (2016). Optimal VAR Control for Real Power Loss Minimization and Voltage Stability Improvement Using Hybrid Multi-Swarm PSO. Proceedings of 2016 IEEE International Conference on Circuit, Power and Computing Technologies [ICCPCT].

34. J. Praveen and Dr. B. Srinivasa Rao. (2016). Multi Objective Optimization for Optimal Power Flow with IPFC Using PSO. Proceedings of 2016 3rd International Conference on Electrical Energy Systems.

35. R. Aggarwal and Y. H. Song. (1997). Fuzzy Logic and Neural Networks in Generation and Distribution. *Power Technology International*, Spring, pp. 39–45.

36. P. K. Dash, T. S. Sidhu, and H. S. Gill. (2000). A Nove lFuzzy Neural Based Distance-Relaying Scheme. *IEEE Transactions Power Delivery*, vol. 15, no. 3, pp. 895–901.

37. Y. H. Song and M. R. Irving. (2001). Optimization Methods for Electric Power Systems, Part 2, Heuristic Optimization Methods. *IEE Power Engineering Journal*, vol. 15, no. 3, pp. 151–160.

38. J. Nanda and R. Badri Narayanan. (2001). Application of Genetic Algorithm to Economic Load Dispatch with Line Flow Constraints. *International Journal of Electrical Power & Energy Systems*, vol. 24, no. 9, November 2002, pp. 723–729.

39. Amita Mahor, Vishnu Prasad, and Saroj Rangnekar. (2009). Economic Dispatch Using Particle Swarm Optimization: Review. *Renewable and Sustainable Energy Reviews*, vol. 13, pp. 2134–2141.

40. R. Hugh, R. Palma, E. Cura, and C. Silva. (1996). Economically Adapted Transmission System in Open Access Schemes—Application of Genetic Algorithm. *IEEE Transaction. Power Systems*, vol. 11, no. 3, pp. 1427–1440.

41. R. Segal, A. Sharma, and M. L. Kothari. (2004). A Self-Tuning Power System Stabilizer Based on Artificial Neural Network. *Electrical Power and Energy Systems*, vol. 26, pp. 423–430.

42. P. Panciatici, M. C. Campi, S. Garatti, S. H. Low, D. K. Molzahn, A. X. Sun, and L. Wehenkel. (2014). Advanced Optimization Methods for Power Systems. 18th Power Systems Computation Conference Wroclaw, Poland—August 18–22.

43. S. Binitha and S. Sathya. (2012). A Survey of Bioinspired Optimization Algorithms. *International Journal of Soft Computing and Engineering (IJSCE)*, vol. 2.

44. Dr. Firas Mohammed Tuaimah and Montather Fadhil Meteb. (2014). A Linear Programming Method Based Optimal Power Flow Problem for Iraqi Extra High Voltage Grid (EHV). *Journal of Engineering*, vol. 20.

45. Jizhong Zho. (2009). *Optimization of Power System Operation*. Institute of Electrical and Electronics Engineers John Wiley & Sons, Inc.

46. Tareq A. Al-Muhawesh and Isa S. Qamber. (2008). The Established Mega Watt Linear Programming Based Optimal Power Flow Model. *Energy*, vol. 33, no. 1, January, pp. 12–21.

47. A. M. Giacomoni and B. F. Wollenberg. (2010). Linear Programming Optimal Power Flow Utilizing. *Saudi Arabia, Elsevier*, vol. 33, no. 1, pp. 12–21.

48. M. G. Morgan and S. Talukdar. (1996). A Trust Region Method, IEEE, North American Power Symposium (NAPS). *Nurturing R&D, IEEE Spectrum*, vol. 33, no. 7, July, pp. 32–33.

49. A. Ye Tao. (2009). Sequential Linear Programming Algorithm for Security-Constrained Optimal Power Flow. IEEE, North American Power Symposium (NAPS) Conference.

50. W. Yamany, A. Tharwat, M. F. Hassanin, T. Gaber, A. E. Hassanien, and T. H. Kim. (2015). A New Multi-layer Perceptrons Trainer Based on Ant Lion Optimization Algorithm. 4th International Conference on Information Science and Industrial Applications (ISI), Busan, pp. 40–45.

51. Guo-Li Zhang, Hai-Yan Lu, Geng-Yin Li, and Guang-Quan Zhang. (2005). Dynamic Economic Load Dispatch Using Hybrid Genetic Algorithm and the Method of Fuzzy Number Ranking. International Conference on Machine Learning and Cybernetics, Guangzhou, China, pp. 2472–2477, vol. 4.

52. K. Asano, M. Nakatsuka, and T. Kumano. (2009). Dynamic Economic Load Dispatch by Calculus of Variation and Genetic Algorithm Considering Ramp Rate. 15th International Conference on Intelligent System Applications to Power Systems, Curitiba, pp. 1–6.

53. C. L. Chiang. (2007). Genetic-Based Algorithm for Power Economic Load Dispatch. *IET Generation, Transmission & Distribution*, vol. 1, no. 2, pp. 261–269.

54. A. Derghal and N. Golea. (2014). Genetic Algorithm for Solving Large Practical Fuzzy Economic Load Dispatch with Prohibited Operating Zones. International Renewable and Sustainable Energy Conference (IRSEC), Ouarzazate, pp. 469–474.

55. J. Eswari and Dr. S. J eyadevi. (2014). An Evolutionary Approach for Optimal Citing and Sizing of Micro-Grid in Radial Distribution Systems. *International Journal of Engineering Trends and Technology (IJETT)*, vol. 11.

56. T. Niknam, M. Narimani, J. Aghaei, S. Tabatabaei, and M. Nayeripour. (2011). Modified Honey Bee Mating Optimisation to Solve Dynamic Optimal Power Flow Considering Generator Constraints. *IET Generation, Transmission & Distribution*, vol. 5, no. 10, pp. 989–100.

57. J. Praveen and B. S. Rao. (2016). *Multi-Objective Optimization for Optimal Power Flow with IPFC Using PSO*. Electrical Energy Systems (ICEES).

Chapter 10

Data Transformation: A Preprocessing Stage in Machine Learning Regression Problems

Akshay Jadhav, Devashish Dhaulakhandi,
Shishir Kumar Shandilya, Lokesh Malviya,
and Arvind Mewada

Contents

DOI: 10.1201/9781003301820-10

10.1 Introduction

In the field of data science and analytics, the data's quantity, features, and properties are the main factors that generate the knowledge base on which decisions and predictions are made. However, data preprocessing and cleaning methods are integral for higher accuracy and a useful outcome before drawing statistical inferences. Raw data is acquired from various sources. It may be quantitative or qualitative in nature, with different degrees of complexity.

For the statistical and machine learning models to successfully infer the trends in the data, it needs to be transformed and scaled into a usable format. The preprocessing or transformation techniques currently are numerous, and so are domains in which machine learning models are implemented.

Machine learning models perform classification [1–3] and regression [4, 5] tasks after learning to identify new predictions based on training data. The learning or training procedure is carried out using the previously known data with a known output value. The data parameters may be categorical, and if they are numerical, they range in scales and units. Some machine learning algorithms and techniques may perform worse than others when learning from raw data. One of the essential steps in the data mining process is the preprocessing of the data, especially the normalization of the data (scaling). This chapter aims to demonstrate how to improve regression problem performance when data normalization is used along with the Random Forest Regression Model.

10.2 Usage of Data Transformation Techniques

It is important to transform the data before modelling to ensure that the machine learning model performs well. It is equally important to select the most optimum transformation technique. Here are some authors and their studies on data preprocessing. Data normalization has been used in various research domains, such as nuclear power plants, medical, biometric systems, credit scores, and many more classification and regression tasks. This section highlights the use of various normalization techniques in different fields.

Sola et al. [6] highlighted data normalization is crucial before neural network training to speed up calculations and produce accurate results for use in nuclear energy generation applications. Jayalakshmi et al. [7] concluded that different transformation techniques enhance the reliability of a trained neural network that uses feed-forward backpropagation. The performance of the classification model of diabetes data using a neural approach also depends on the normalization methods. Zhang et al. [8] offered a normalization technique considered appropriate for the specific data set from the University of California Irvine (UCI) repository. However, since the results are not universally applicable, the need for an optimum transformation remains.

Jain et al. [9] reported that MinMax and Z-score methods are better than other normalization methods and play a vital role when used with biometric systems. With credit score data, Huang et al. [10] employed the MinMax normalization to scale the characteristics in the [10, 11] range. Wang et al. [11] used evolutionary algorithm-based feature selection to improve the classification of credit approval data and used MinMax normalization for preprocessing. Li et al. [12] enhanced the categorization of intrusion data using SVM and tested several normalization techniques.

Kadir et al. [13] described the leaf classification, a Probabilistic Neural Network (PNN) was utilized, and the retrieved features were scaled using the min-max approach. Pan et al. [14] examined the impact of various data normalization techniques for forecasting stock indexes. Malviya et al. [15] used a standard scaling technique for normalizing the EEG data for stress detection on subjects using a hybrid deep learning model. Kardani et al. [16] employed the MinMax scaling technique to normalize the whole data set to predict the permeability of tight carbonates.

10.3 Methodology

10.3.1 Transformation Techniques

The transformation techniques we used for preprocessing are MinMax Scaling, Standard Scaling, Max Absolute Scaling, Quantile Transformation, Yeo–Johnson Transformation, and Robust Scaling [17, 18].

10.3.1.1 MinMax Scaler

It transforms features by scaling each feature to a specific range. This estimator scales and translates each feature separately so that it falls inside the training set's predetermined range, which is often regarded as between zero and one, provided by the Equation 10.1.

$$x_{Scaled} = \frac{(x_i - \min(x))}{\max(x) - \min(x)} \tag{10.1}$$

10.3.1.2 Standard Scaler

It uniformizes the features by reducing the mean and scaling to unit variance. The standard score of a sample x is calculated using Equation 10.2.

$$x_{Scaled} = \frac{(x_i - u)}{s} \tag{10.2}$$

where, s is the training samples' standard deviation, and u is the training samples' mean. Through the computation of the pertinent statistics on the samples in the training set, cantering and scaling occur independently on each feature. Mean and standard deviation are then stored to be used on later data.

10.3.1.3 MaxAbs Scaler

The MaxAbs scaler scales each feature based on its highest absolute value in the training set. The scaling is done separately for each feature, and the maximum absolute value of each feature in the training set is set to 1.0. This scaler does not shift or center the data, so sparsity is preserved. The MaxAbs scaler is calculated using Equation 10.3.

$$x_{Scaled} = \frac{(x_i)}{\max(x)} \tag{10.3}$$

10.3.1.4 Quantile Transformation

The quantile transformation technique adds quantile-based scaling, which is applied individually to each feature. The first step is to map the original feature values to a uniform distribution by estimating its cumulative distribution function. The quantile function is then used to transform the uniform distribution to the desired output distribution, which is calculated using Equation 10.4.

$$Q(p) = \inf \{x \in R : p \leqslant F(x)\} \tag{10.4}$$

where, the quantile function Q returns a threshold value is x from among all those values whose cumulative

Equation 10.5 represents the inverse of the cumulative distribution function, which maps values from [0,1] to the original feature distribution. This is useful for transforming the features to have a uniform or normal distribution, where values are modified so that they do not exceed a certain probability p.

$$F_{x:} R \to [0,1] \tag{10.5}$$

10.3.1.5 Yeo–Johnson Transformation

The features are transformed to follow a uniform or normal distribution, a continuous probability curve for a random variable, to make the data more Gaussian-like. The Gaussian distribution's generalized equation is represented by Equation 10.6.

$$f(x) = \frac{1}{\sigma\sqrt{2\pi}} e^{\left(-\frac{1}{2}\left(\frac{x-\mu}{\sigma}\right)^2\right)} \qquad (10.6)$$

where parameter μ is the mean, σ is its standard deviation.

This is helpful when modeling heteroscedasticity (nonconstant variance) problems or other circumstances when normalcy is preferred. The ideal parameter for reducing skewness and stabilizing variance is determined using maximum likelihood. The Yeo–Johnson Transformation can support both positive and negative data. The processed data is then normalized with a zero-mean, unit-variance distribution.

10.3.1.6 Robust Scaler

The Robust Scaler scales features using statistics that are robust to outliers, based on the interquartile range. This scaler computes relevant statistics on the training set samples and scales and centers each feature individually. The median and interquartile ranges are saved for use in future data. The calculations are performed using Equation 10.7.

$$x_{Scaled} = \frac{(x_i - x_{med})}{x_{Q3} - x_{Q2}} \qquad (10.7)$$

10.3.2 Data Set Description

The chapter deals with regression-type problems from multiple domains. A total of six data sets are utilized in the chapter, sourced from various fields like medical data, stocks data, data of freshwater, marine life, runtime data of GPUs, energy consumption and efficiency, the volume of traffic in a metro-city, software effort estimation, and intensity of forest fires. The data sets vary in the number of features, entities, and sizes. This chapter aims to identify a data preprocessing technique suitable for any data size despite the number of features and entities. Another objective is to observe the impact of these transformation techniques when applied to the machine learning model. The detailed data set, a description with the number of observations, features, and target field in each data set, along with their mean and median, is well-represented in Table 10.1.

Table 10.1 A Detailed Description of Regression Data Sets

Data Sets	#Features	#Entities	Target Field	Mean	Median
Maxwell [21]	28	62	Effort	8223.21	5189.50
Fish Weights [22]	5	159	Weights	398.33	273.00
Insurance [23]	6	1338	Charges	13270.42	9382.03
Netflix [24]	5	1009	Volume	7570685.04	5934500.00
GPU Runtimes [25]	18	241600	Run4	217.53	69.82
Energy Efficiency [26]	8	768	Load	24.59	22.08
Forest Fires [27]	13	517	Area Affected	12.85	0.52
Metro-City Traffic [28]	9	48204	Traffic Volume	3259.82	3380.00

10.3.3 Machine Learning Model

The Random Forest [19, 20] machine learning regression model is utilized as a supervised learning technique for leveraging the ensemble learning approach for regression. The ensemble approach combines predictions after fitting numerous classification decision trees to reduce overfitting and provide more accurate predictions than those from a single decision tree. Random Forest is classified as a highly robust and precise regression model, usually performing great on many problems, including non-linear relationships. The accuracy of the Random Forest Regressor is directly affected by the quality of the data used in training feature variables. Different scales do not contribute generously to the model fitting and learned function and create a bias. In the scope of this chapter, we observe the effects of data belonging to different domains, transformed using various available techniques, on the accuracy of the Random Forest Regressor.

10.4 Proposed Approach

The introduction section discusses the importance of transformation or preprocessing techniques. It leads us to the assumption that there has to be some technique that is robust to the outliers in the data and compatible with different quantities and qualities of data (small, medium, or large, number of features, ranges of feature values), and simple to implement. This section examined the currently used data transformation methods mentioned in the previous sections. We studied their outcomes on different data sets from various fields belonging to regression tasks. The flow of the experiment is shown in Figure 10.1. The data sets were loaded one by one and divided into train and test split ratio of 70:30 such that 70% of the data was transformed according to the preprocessing technique and then used to train the Random Forest Regressor Model with the default parameters [n estimators = 100, random _ state = 0]. The remaining 30% of the data was then separately used to test the model's accuracy. To prove the effectiveness and applicability of the techniques, we have constructed a machine-learning regression model pipeline using a Random Forest Regressor and evaluated the regression accuracy metric of R-square value (R^2) for better comprehension after performing sample validation using k-fold cross-validation. The description of the comparison research is provided in the next section using data tabulation and graphical depiction.

10.5 Results and Discussions

The proposed study investigates how data normalization techniques influence regression performance. The entire collection of features is considered while evaluating these normalization techniques. In this study, eight regression datasets were chosen

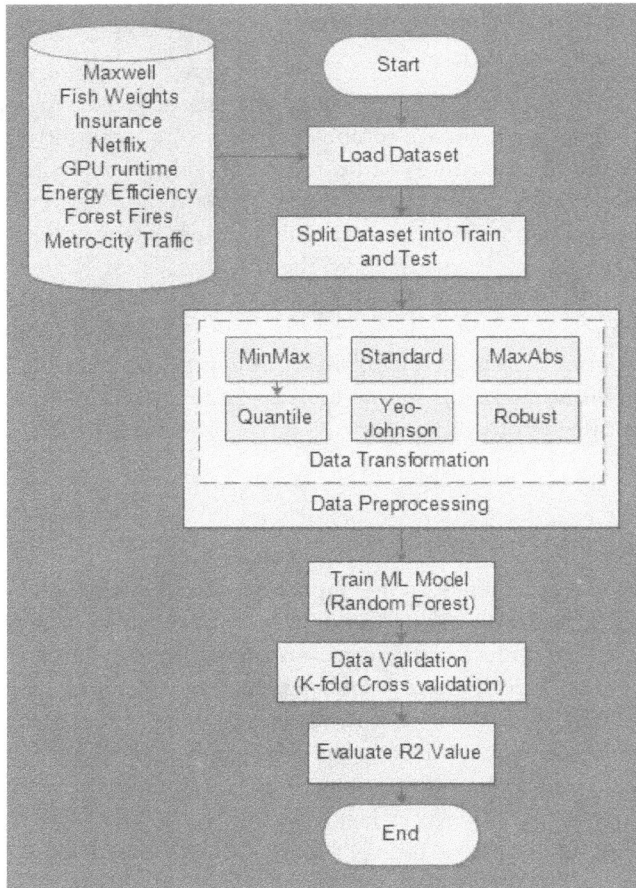

Figure 10.1   Proposed Model Architecture.

from different fields, varying in the number of features, observations, and data size. This section shows the performance of these data sets after applying all six data transformation techniques with the Random Forest model. The comparison of the impact of these techniques with respect to evaluation metrics over the Random Forest Regressor is shown in Table 10.2.

Figure 10.2 represents the graphical comparison of R-squared values obtained after each transformation technique, applied to the chosen eight data sets and the Random Forest Regressor model is trained and validated. The figure consists of eight bar graphs showing the performance of each data transformation technique on each dataset when trained using a Random Forest Regression model. The result shows the

Table 10.2   Statistical Comparison of Transformation Techniques on Different Data Sets

Data sets	TT1	TT2	TT3	TT4	TT5	TT6
Maxwell	0.653	0.663	0.645	0.645	0.685	0.651
Fish Weights	0.859	0.859	0.859	0.859	0.859	0.858
Insurance	0.840	0.840	0.841	0.840	0.840	0.840
Netflix	0.818	0.818	0.818	0.818	0.818	0.818
GPU Runtimes	0.994	0.994	0.994	0.994	0.994	0.994
Energy Efficiency	0.980	0.980	0.980	0.980	0.980	0.980
Forest Fires	0.910	0.910	0.913	0.913	0.913	0.910
Metro-City Traffic	1.000	1.000	1.000	1.000	1.000	1.000
TT1: MinMax Scaler, TT2: standard scaler, TT3: Max-Abs Scaler TT4: quantile transformation, TT5: Yeo–Johnson Transformation, TT6: robust scaling						

R^2 value obtained after training in most data sets is maximum when Yeo–Johnson data transformation is applied to the data sets. Only in the insurance and forest fires data set the value obtained by Yeo–Johnson is slightly less than or similar to the Max-Abs Scaling Technique.

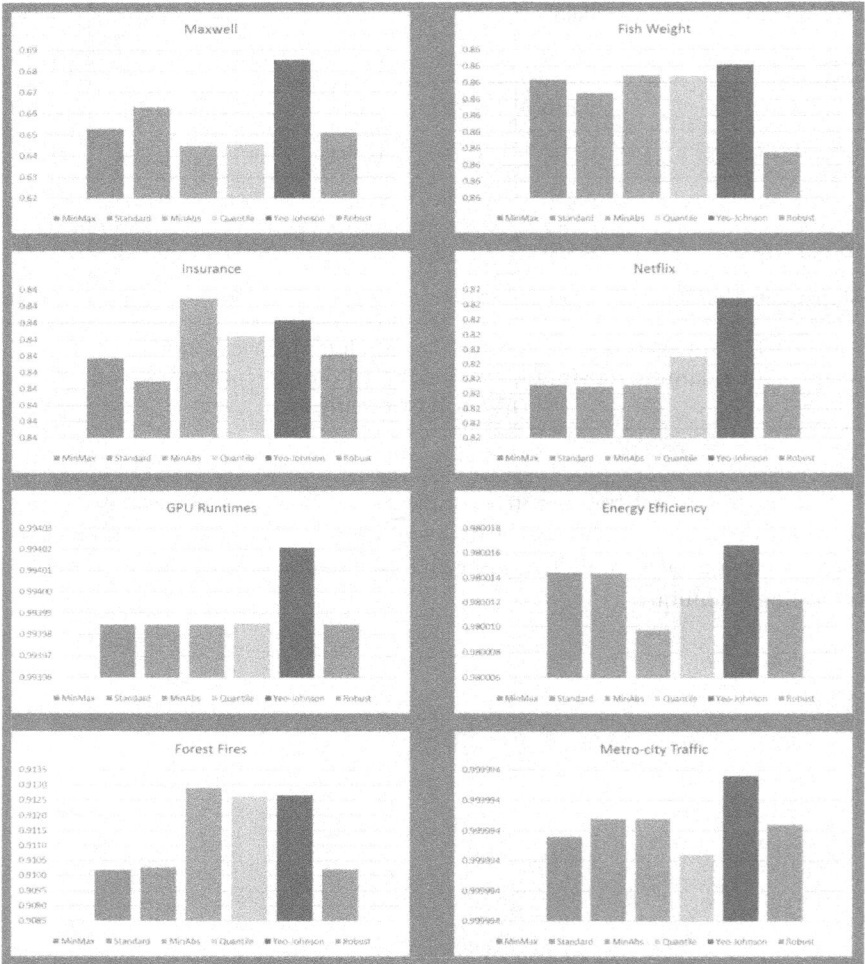

Figure 10.2 Comparison of Obtained R-Squared Value on Different Data Sets.

10.6 Conclusion and Future Work

Preparing raw data to be acceptable for a machine learning model is called data preprocessing. Raw data typically includes noise and missing values and may be unfavorable, making it impossible to create machine learning models on it directly. The preprocessing is divided into four stages: cleaning, integration, reduction, and transformation. The paper deals with data transformation techniques and identifies the best suitable and stable approach for data normalization. On performing certain experiments on different regression data sets problems, the chapter identifies Yeo–Johnson Transformation as suitable with the maximum number of regression data sets, followed by Max Absolute Scaling. Both these techniques outperform other transformation techniques. In the future, we intend to propose a hybrid transformation technique combining these two techniques to improve the performance of machine learning and deep learning models for regression and classification problems.

References

1. D. Singh and B. Singh, "Investigating the impact of data normalisation on classification performance," *Applied Soft Computing*, vol. 97, p. 105524, 2020.
2. H.-C. Huang and L.-X. Qin, "Empirical evaluation of data normalisation methods for molecular classification," *PeerJ*, vol. 6, p. e4584, 2018.
3. E. Basan, A. Basan, A. Nekrasov, C. Fidge, E. Abramov, and A. Basyuk, "A data normalisation technique for detecting cyber-attacks on UAVS," *Drones*, vol. 6, no. 9, p. 245, 2022.
4. S. Banik, N. Sharma, and K. Sharma, "Analysis of regression techniques for stock market prediction: A performance review," in *2021 9th International Conference on Reliability, Infocom Technologies and Optimization (Trends and Future Directions) (ICRITO)*. IEEE, 2021, pp. 1–5.
5. D. H. Fudholi, R. A. N. Nayoan, and S. Rani, "Stock prediction based on Twitter sentiment extraction using BiLSTM-attention," *Indonesian Journal of Electrical Engineering and Informatics (IJEEI)*, vol. 10, no. 1, pp. 187–198, 2022.
6. J. Sola and J. Sevilla, "Importance of input data normalisation for the application of neural networks to complex industrial problems," *IEEE Transactions on Nuclear Science*, vol. 44, no. 3, pp. 1464–1468, 1997.
7. T. Jayalakshmi and A. Santha Kumaran, "Statistical normalisation and backpropagation for classification," *International Journal of Computer Theory and Engineering*, vol. 3, no. 1, pp. 1793–8201, 2011.
8. Q. Zhang and S. Sun, "Weighted data normalisation based on eigenvalues for artificial neural network classification," in *International Conference on Neural Information Processing*. Springer, 2009, pp. 349–356.
9. A. Jain, K. Nandakumar, and A. Ross, "Score normalisation in multimodal biometric systems," *Pattern Recognition*, vol. 38, no. 12, pp. 2270–2285, 2005.
10. C.-L. Huang and J.-F. Dun, "A distributed PSO–SVM hybrid system with feature selection and parameter optimisation," *Applied Soft Computing*, vol. 8, no. 4, pp. 1381–1391, 2008.

11. C.-M. Wang and Y.-F. Huang, "Evolutionary-based feature selection approaches with new criteria for data mining: A case study of credit approval data," *Expert Systems with Applications*, vol. 36, no. 3, pp. 5900–5908, 2009.

12. Z. Liu *et al.*, "A method of SVM with normalisation in intrusion detection," *Procedia Environmental Sciences*, vol. 11, pp. 256–262, 2011.

13. A. Kadir, L. E. Nugroho, A. Susanto, and P. I. Santosa, "Leaf classification using shape, colour, and texture features," *arXiv preprint arXiv:1401.4447*, 2013.

14. J. Pan, Y. Zhuang, and S. Fong, "The impact of data normalisation on stock market prediction: Using SVM and technical indicators," in *International Conference on Soft Computing in Data Science*. Springer, 2016, pp. 72–88.

15. L. Malviya and S. Mal, "A novel technique for stress detection from EEG signal using hybrid deep learning model," *Neural Computing and Applications*, vol. 197, pp. 1–12, 2022.

16. N. Kardani, A. Bardhan, B. Roy, P. Samui, M. Nazem, D. J. Armaghani, and A. Zhou, "A novel improved Harris hawks optimisation algorithm coupled with elm for predicting permeability of tight carbonates," *Engineering with Computers*, vol. 38, pp. 1–24, 2021.

17. P. Huilgol, "9 feature transformation & scaling techniques—boost model performance," 2020, www.analyticsvidhya.com/blog/2020/07/types-of-featuretransformation-and-scaling/ [Accessed 19-May-2022].

18. L. Buitinck, G. Louppe, M. Blondel, F. Pedregosa, A. Mueller, O. Grisel, V. Niculae, P. Prettenhofer, A. Gramfort, J. Grobler, R. Layton, J. VanderPlas, A. Joly, B. Holt, and G. Varoquaux, "API design for machine learning software: Experiences from the Scikit-learn project," in *ECML PKDD Workshop: Languages for Data Mining and Machine Learning*. CoRR, 2013, pp. 108–122.

19. I. Reis, D. Baron, and S. Shahaf, "Probabilistic Random Forest: A machine learning algorithm for noisy datasets," *The Astronomical Journal*, vol. 157, no. 1, p. 16, 2018.

20. E. Scornet, "Tuning parameters in random forests," *ESAIM: Proceedings and Surveys*, vol. 60, pp. 144–162, 2017.

21. Y. Li, "Effort estimation: Maxwell," March 2009, https://doi.org /10.5281/zenodo.268461.

22. A. Pyae, "Fish market," June 2019, www.kaggle.com /datasets/aungpyaeap/fish-market.

23. "Medical cost personal datasets—kaggle.com," www.kaggle.com/datasets/mirichoi 0218 /insurance [Accessed 11-June-2022].

24. "Netflix stock prices dataset—data science and machine learning—kaggle.com," www.kaggle.com/general/168472 [Accessed 15-June-2022].

25. R. Ballester-Ripoll, E. G. Paredes, and R. Pajarola, "Sobol tensor trains for global sensitivity analysis," 2017, https://arxiv.org/abs/1712.00233.

26. A. Tsanas and A. Xifara, "Accurate quantitative estimation of the energy performance of residential buildings using statistical machine learning tools," *Energy and Buildings*, vol. 49, pp. 560–567, 2012.

27. NEVES, José Maia; SANTOS, Manuel Filipe; MACHADO, José Manuel, eds. - "*New trends in artificial intelligence: proceedings of the 13th Portuguese Conference on Artificial Intelligence* (EPIA 2007), Guimarães, Portugal, 2007." APPIA, 2007. ISBN 978-989-95618-0-9. p. 512–523.

28. D. Dua and C. Graff, "UCI machine learning repository," 2017, https://archive.ics.uci.edu /ml /datasets/Metro+Interstate+Traffic+Volume.

Chapter 11

Predicting Native Language with Machine Learning: An Automated Approach

Dasangam Venkat Nikhil, Rupesh Kumar Dewang,
Buvaneish Sundar, Ayush Agrawal,
Ananta Narayan Shrestha, Akash Tiwari,
and Arvind Mewada

Contents

DOI: 10.1201/9781003301820-11

11.1 Introduction

The identification of the native language (l1) of an author based on their writings in another language (l2) is called Native Language Identification [1]. Native Language Identification (NLI) has been increasingly important in natural language processing and machine learning communities. English is the most widely used second language in the world. It is predominantly used in social media, online discussion forums, blogs, and posts, but few sharp contrasts exist in how people from different countries learn and speak the language. For example, Chinese speakers generally tend to use long sentences than others. These features can capture the origin of the speaker/writer based on their writing.

It paves the avenue for state of the art in crime detection. We can also identify a user's geographical location in a discussion group or a forum without explicitly asking them. NLI is also extensively used in Second Language Acquisition, in which how people learn a second language with the help of their first language is studied. NLI generally acts as the first step toward the larger goal of author profiling [2], in which the characteristics of an author are determined based on their writing.

Our approach is partially based on "Experimental results on the native language identification shared task" [3]. In this paper, the authors have taken 9,900 sample essays, considering speakers from 11 different countries, using a support vector machine classifier. They obtained an initial accuracy of 43%, and with improved feature normalization, they obtained an accuracy of 63%. On the other hand, we have considered data from the International Corpus Network of Asian Learners of English (ICNALE). We have considered predominantly Asian countries. It is challenging because Asian languages are linguistically highly similar. We have implemented logistic regression and artificial neural networks by varying the hyperparameters and have obtained an average accuracy of 76.56% and 73.13%, respectively.

11.2 Related Work

Most research in Native Language Identification has been done either using lexical features, syntactic features, or a combination of both. The amount of recognition this field has received is primarily due to the NLI Shared Task 2013. The lexical approach generally uses features like character or word N-grams, stemming, and support vector machines, which have been the most successful in the lexical approach.

The syntactic approach generally uses parse trees, dependency trees, and part-of-speech tagging. Increasingly, ensemble classification approaches have yielded better results and are used extensively [4]. Malmasi et al. [5] gave a detailed description of NLI research during 2001–2015, in which the first work on this field was done by Koppel et al. [4], where they explored many features, including POS n-grams, content and feature words, as well as spelling and grammatical errors. A support vector machine (SVM) model was trained on these features extracted from a subsection of the first version of the International Corpus of Learner English (ICLE), consisting of writers from five (Bulgarian, Czech, French, Russian, or Spanish) different native languages. The existing research has explored different N-gram features (word, character, and POS), in which most of the research focuses on unigrams and bigrams. Besides being relatively easy to compute, N-gram features also have a high accuracy rate. They mostly explored the ungrammatical structures with rare POS bigrams from the Brown corpus. Koppel et al. [1] and Tsur and Rappoport [6] have addressed the significance of word choices in second-language writing. They noted that the frequency of native language vocable highly influences the choice of words used in English writing. Estival et al. [7] tackled the work of identifying and exploring author profiles across a smaller set of languages (English, Spanish, and Arabic). Wong and Dras [2] explored the versatility of syntactic features using the top 200 PoS bigrams to identify errors using statistical parsers, and further research by them in 2012 explored the usage of Adaptor Grammars in identifying many useful features using a mixture of PoS tags and Function Words. Tsur and Rappoport [6] highlighted the significance of phonemes that can be explored through character N-grams in writing, capturing different lexical, punctuational usage, and capitalization schemes. Their work explored how phonemes can be used in the prospect of figuring out a second-language writer's native language.

Moreover, Swanson and Charniak [8] and Tetreault et al. [4] explored the utility of Tree Substitution Grammars (TSGs) using dependency features from Stanford Parser. Brooke and Hirst [9] explored and illustrated the effects of training/testing on various corpora. In their paper "Identifying the L1 of non-native writers: the CMU-Haifa system," Tsvetkov et al. [10] used logistic regression to predict the most probable native language from a set of 11 possibilities with high accuracy.

Similarly, Jervis et al. [11] investigated a more considerable diversity of errors and their significance in identifying the native language of the writer. In conclusion, various aspects can be explored in Native Language Identification, precisely as lexical and syntactic information, structural information, idiosyncrasies, and errors.

11.3 Proposed Approach

We have implemented a feature-based classification approach using data primarily from Asian countries due to the linguistic similarities among Asian languages. The NLI task becomes even more challenging as a result.

11.3.1 Proposed Model

The first step of the experiment involved data collection, which was carried out by gathering information from a single source, as described in the following section. The subsequent steps involved feature extraction, normalization, and implementation of two algorithms, namely logistic regression and artificial neural networks, as explained in the later sections. Finally, the obtained results were compared and evaluated, as shown in Figure 11.1.

11.3.2 Dataset Description

The training and testing data sets are from the International Corpus Network of Asian Learners of English (ICNALE). This dataset comprises 5,600 essays written on a selected set of topics. The people writing these essays are divided into ten categories: China, Japan, Korea, Singapore, Philippines, Indonesia, Taiwan, Pakistan, Thailand, and Hongkong. One more category is used for essays written by native speakers of English, like those in Australia and England. The dataset has been divided into training and testing sets in the ratio of 80: 20. This is done using the cross-validation split in Python to ensure a regularized split.

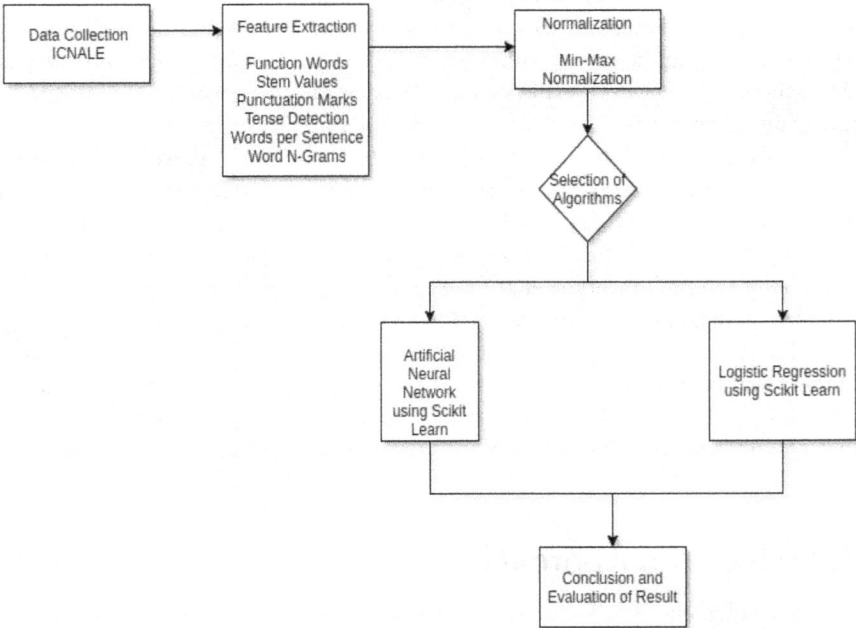

Figure 11.1   Flow Diagram of the Model.

11.3.3 Features [3]

The features utilized in this experiment are categorized into six distinct categories, including Function Words, comprising a total of 319 features, Stem Values, Words per Sentence, and Punctuation Marks, which consists of two features. The feature set also includes Tense as a Feature, which includes three distinct features—one for each tense, past, present, and future. Additionally, N-Grams, a feature set comprising 200 features, is also included in the experiment.

11.3.3.1 Function Words

Function words barely possess any meaning but have syntactic significance in the sentence. They are just used to make the sentence grammatically complete and consistent. They're moreover known as grammatical words or empty words. The usage of these words in the writing reveals a great deal about the native language of the author. The English language has a total of 319 function words. Examples of a few function words are "AGAIN,""AGO," "ALMOST,""ALREADY," "ALSO,""ALWAYS,""ANYWHERE,""BACK,""ELSE,""EVEN,""EVER," and "EVERYWHERE." For every document, we counted the number of occurrences of each function word present, thus creating a vector of 319 elements to be used as features.

11.3.3.2 Stem Values

The stem of a word refers to the underlying word from which the specific word is derived. English speakers from diverse regions have varying degrees of vocabulary usage. The word stems utilized in written works can reveal the author's language proficiency and diversity. Authors from different native languages tend to use varying degrees of vocabulary, which can be captured by counting the unique number of stems as an additional feature in identifying the native language. Some examples of word stems include playful (play), partnership (partner), and running (run).

11.3.3.3 Words per Sentence

Speakers of English from different countries differ in the number of words used in a sentence. For example, Chinese speakers use longer sentences than others when they write. The number of words per sentence has been calculated for each document and used as a feature. For example, consider two sentences like "This is our institute project." and "This project is submitted in the seventh semester." have a total number of sentences = 2, in which a total number of words = 13, and the average number of words per sentence is 7.5.

11.3.3.4 Punctuation Marks

Punctuation marks help a great deal in identifying the writer's original, native language. Since the writer tries to replicate that aspect of the native language into English while learning and writing, it's interesting to screen the possibilities of influence of it on writing. Some communities could use more commas (",") in place of full stops ("."), while others could prefer using semicolons (";"). This way, Native Language Identification can be made using punctuation marks. There are around 17 punctuation marks in the English language. Examples of punctuation marks are ".","","","," and "!". To capture these kinds of patterns, two things have been considered: the number of punctuation marks per word and the number of punctuation marks per sentence. These two have been computed for each document to be used as two separate features.

11.3.3.5 Tense as a Feature

To understand more information about the native language of an author, the number of sentences in present, past, and future tenses in a document can be used as three separate features. To compute this, the part-of-speech tagging method from Python's Natural Language Toolkit (NLTK) library has been used. For every sentence, we consider the number of verbs of each present tense to determine the tense of the entire sentence. Then, for each document, we determine the number of sentences in past, present, and future tenses. For example, consider the two sentences "I am going there." and "I had gone there." In this context, the first sentence is in the present tense, and the second one is in the past tense. So, for this document, the number of sentences in the past tense is 1, the number of sentences in the present tense is 1, and the number of sentences in the future tense is 0. Using this data from every document, we consider these as three separate features.

11.3.3.6 Word N-Grams

An N-gram is a sequential combination of all adjoining words, letters, or n-length tokens. Word N-grams means a combination of all words from the text source adjacent to each other in an N-word sequence. The words in a sentence generally correlate with the position of words in the sentence. The position of words in a sentence plays a massive role in identifying the syntactic and grammatical relationship of words in a sentence. Different variations of the N-gram include unigram, bigram, and trigram, depending on the number of words in sequence used to form an N-gram. Unigram selects a single word at a time, while bigram illustrates the grouping of two words in exploring the sentence accordingly. For example, consider the sentence, "I am doing my project." Then unigrams would be a list of words [I, am, doing, my, project], and bigrams would be a list of tuples like [(I, am), (am, doing), (doing, my), (my, project)]. We have considered only bigrams for this problem. First, we traversed through every document, formed a list of the top 200 most used bigrams, and used each as a feature. For each document, we have calculated the number of occurrences of each of these selected bigrams, thus resulting in 200 features.

11.4 Algorithms

A logistic regression model in the Scikit-learn module of Python and an artificial neural network were used for classification.

11.4.1 Logistic Regression

Logistic regression is a supervised learning algorithm that uses the logistic function as its key element. This function is also known as the sigmoid function, which maps the input values to a range between 0 and 1.

The hypothesis function of logistic regression is obtained by applying the logistic function to the linear combination of the input features and their corresponding parameters. The algorithm calculates these parameters during the training process using optimization techniques such as the gradient descent algorithm. The logistic function is mathematically represented by Equation 11.1.

$$Logit(p) = \frac{1}{(1 + (e^{-p})}$$ (11.1)

In this experiment, the one-versus-rest model is used for Native Language Identification. We have a total of 11 classes in our problem. In this model, for each of the 11 classes, the probability that the given writing belongs to the particular class is calculated. Then, for whichever class we obtain the maximum accuracy, the given writing is mapped to that particular class. For the problem in the context (Native Language Identification), for the implementation, we have used the logistic regression model in the Scikit-learn module of Python. First, we obtain a model classifier and use the fit method to train the data based on the training set. Then, we use the scoring method on the testing dataset to obtain the trained classifier's accuracy.

The tolerance is a condition that can be set to stop further iterations in the algorithm. To control overfitting, the inverse of the regularization strength is used, with smaller values indicating more significant regularization. The logistic regression classifier utilized the Liblinear solver with a maximum of 100 iterations of the gradient descent algorithm, resulting in an average final accuracy of 76.56%, as shown in Table 11.1.

11.4.2 Artificial Neural Network

An artificial neural network is a computation system inspired by biological neural networks. It consists of input, output, and hidden layers. Each layer consists of

Table 11.1   Logistic Regression Result

Parameter	Value
Tolerance	0.0001
Regularization Strength C	1
Max-iter (No. of iterations)	100
Multi-class (Model used)	ovr

neurons that are fully connected to the next layer. Each neuron acts as a sigmoid neuron and implements a logistic function described next. Like their counterpart perceptron, sigmoid neurons have input for different weights (w_1, w_2) and bias b. They hold significant distinction to have output values ranging from 0 to 1 and then holding to 0 or 1, and the output is given in Equation 11.2.

$$\sigma(w_x + b), \tag{11.2}$$

Here the sigmoid function is given by Equation 11.3, σ

$$\sigma(z) = \frac{1}{1 + e^{-z}} \tag{11.3}$$

Having inputs as x_1, x_2,. . ., x_n weights as w_1, w_2,. . ., w_n and the following equation gives bias as b, the output of sigmoid neuron given in Equation 11.4.

$$\frac{1}{1 + exp(-\sum_j w_j x_j - b))} \tag{11.4}$$

In the algorithm, the probability of the input feature vector being mapped to each class value is calculated, and the class with the maximum probability is the class to which the input given is mapped. For the problem in the context (Native Language Identification), the Multi-Layer Perceptron Classifier (MLP Classifier) model in the Scikit-learn module of Python was used. First, a model classifier is obtained, and the appropriate method is used to train the data based on the training set. Then, the scoring method is used on the testing data set to obtain the trained classifier's accuracy. The number of hidden layers, learning rate, layer sizes (number of neurons in each layer), mini-batch size, and epochs (number of iterations) is known as hyperparameters. By varying the hyperparameters, the accuracy varies. In this project, the input layer contains all the input features, so the input layer has a total of 525 features. The output layer has 11 neurons corresponding to 11 languages. For a given test case, these output neurons each have a value between 0 and 1.

Table 11.2   ANN Hyperparameters

S. No.	Hyperparameters	Value
1	Number of hidden layers	2
2	Hidden layer size of each hidden layer	200
3	Tolerance factor	0.0001
4	Constant learning rate	0.01
5	Maximum number of iterations	300
6	Batch size	500

Table 11.3   Number of Hidden Layers and Change in the Corresponding Accuracy

#HL	Accuracy
1	71.69%
2	72.67%
3	70.00%
4	70.44%
5	72.76%
#HL=number of hidden layers	

The feature vector will be assigned a class that has the maximum probability. The hyperparameter list and its values are given in Table 11.2. The solver was chosen to be "sgd," which stands for stochastic gradient descent, in which there are six varied parameters.

The tolerance factor refers to the stopping condition of the gradient descent algorithm. The learning rate of gradient descent is chosen as a constant value. The other parameters varied are the maximum number of iterations per instance of running the gradient descent, number of hidden layers, number of neurons per hidden layer, and batch size for implementing stochastic gradient descent. After executing the algorithm with the previously mentioned hyperparameters, we obtained an average final accuracy of 73.13%. We have tried to further increase the accuracy by varying the hyperparameters. The corresponding changes in the accuracy with the changes in the hyperparameters are depicted in Table 11.3.

Table 11.4   Mini Batch Size and Change in the Corresponding Accuracy

MBS	Accuracy
500	72.76%
1000	72.41%
1500	70.89%
2000	72.32%
2500	69.10%
MBS=mini batch size	

Table 11.5   No. of Iterations and Change in the Corresponding Accuracy

#Iterations	Accuracy
200	70.15 %
300	71.17 %
400	72.51 %
500	70.87 %
600	71.51 %

The accuracy changes inconsistently with the increase in the number of hidden layers, suggesting no direct or linear relationship between the number of hidden layers and accuracy.

The accuracy of the model changes in an inconsistent manner with the increase in the batch size used in the stochastic gradient descent. This observation suggests that there is no direct or linear relationship between the batch size and accuracy, as shown in Table 11.4.

The relationship between the number of iterations and accuracy is not direct or linear, as shown in Table 11.5. The accuracy changes inconsistently with an increase in the number of iterations in stochastic gradient descent.

Table 11.6 shows that there is no clear linear relationship between the number of neurons per hidden layer and the resulting accuracy. The observed changes in accuracy are inconsistent with the increase in the number of neurons per hidden layer.

Table 11.7 presents a performance comparison of the proposed model using two different algorithms: logistic regression and artificial neural network. The table shows the accuracy achieved by each algorithm in percentage. According to the table, the logistic regression algorithm achieved an accuracy of 76.56%, while the artificial neural network algorithm achieved an accuracy of 73.13%.

Table 11.6   No. of Neurons and Change in the Corresponding Accuracy

#Neurons	Accuracy
500	72.86 %
1000	72.95 %
1500	70.12 %
2000	71.65 %
2500	70.10 %

Table 11.7   Performance Comparison of the Proposed Model

Algorithm	Accuracy
Logistic Regression	76.56 %
Artificial Neural Network	73.13 %

It is interesting to note that the accuracy achieved by each algorithm does not show any direct or linear relationship with the hyperparameters used in the model. This implies that simply increasing or decreasing the values of hyperparameters may not lead to improved accuracy. Instead, optimizing the algorithm by trying out all possible values using Grid Search CV or genetic algorithms may be necessary to achieve the best possible accuracy.

11.5 Conclusion

An accuracy of 76.56% was obtained using the logistic regression classifier, and 73.13% accuracy was achieved using artificial neural networks (refer to Table 11.7). Compared to "Experimental results on the native language identification shared task" [3], a support vector machine classifier was used to achieve an accuracy of 43%, which improved to 63% with enhanced feature normalization. This suggests that features such as N-Grams, Function Words, and Tense Detection can effectively model the problem statement, even for highly similar languages. When comparing the two algorithms used in our study, logistic regression performed slightly better than ANN.

11.6 Future Work

The work can be further improved by considering a larger dataset. For this study, 5,600 essays from primarily Asian countries were used for training and testing purposes. However, the scope can be extended to non-English speaking countries outside Asia, such as France, Germany, Spain, and Argentina, by collecting additional datasets. Logistic regression and neural network models were implemented in this project, but it can be further improved by using deep neural networks (DNNs) and SVMs and integrating speech-based NLI with text-based NLI. There are two primary speech-based NLI approaches, acoustic and phonotactic features, that can be implemented independently or integrated with text-based identification methods.

References

1. M. Koppel, J. Schler, and S. Argamon, "Computational methods in authorship attribution," *Journal of the Association for Information Science and Technology*, vol. 60, no. 1, pp. 9–26, 2009.
2. S.-M. J. Wong and M. Dras, "Exploiting parse structures for native language identification," in *Proceedings of the Conference on Empirical Methods in Natural Language Processing*. Association for Computational Linguistics, 2011, pp. 1600–1610.
3. A. Abu-Jbara, R. Jha, E. Morley, and D. Radev, "Experimental results on the native language identification shared task," in *Proceedings of the Eighth Workshop on Innovative Use of NLP for Building Educational Applications*, 2013, pp. 82–88.
4. J. Tetreault, D. Blanchard, and A. Cahill, "A report on the first native language identification shared task," in *Proceedings of the Eighth Workshop on Innovative Use of NLP for Building Educational Applications*, 2013, pp. 48–57.
5. S. Malmasi, K. Evanini, A. Cahill, J. Tetreault, R. Pugh, C. Hamill, D. Napolitano, and Y. Qian, "A report on the 2017 native language identification shared task," in *Proceedings of the 12th Workshop on Innovative Use of NLP for Building Educational Applications*, 2017, pp. 62–75.
6. O. Tsur and A. Rappoport, "Using classifier features for studying the effect of native language on the choice of written second language words," in *Proceedings of the Workshop on Cognitive Aspects of Computational Language Acquisition*. Association for Computational Linguistics, 2007, pp. 9–16.
7. Dominique Estival, Tanja Gaustad, Son Bao Pham, Will Radford, and Ben Hutchinson, "Author profiling for English emails," in *Proceedings of the 10th Conference of the Pacific Association for Computational Linguistics*, vol. 263, 2007, p. 272.
8. Ben Swanson and Eugene Charniak, "Native language detection with tree substitution grammars," in *Proceedings of the 50th Annual Meeting of the Association for Computational Linguistics (Volume 2: Short Papers)*, 2012, pp. 193–197.
9. Julian Brooke and Graeme Hirst, "Native language detection with 'cheap' learner corpora," in *Twenty Years of Learner Corpus Research. Looking Back, Moving Ahead: Proceedings of the First Learner Corpus Research Conference (LCR 2011)*, vol. 1. Presses universitaires de Louvain, 2013, p. 37.
10. Y. Tsvetkov, N. Twitter, N. Schneider, N. Ordan, M. Faruqui, V. Chahuneau, S. Wintner, and C. Dyer, "Identifying the l1 of non-native writers: The CMUHAIFA system," in *Proceedings of the Eighth Workshop on Innovative Use of NLP for Building Educational Applications*, 2013, pp. 279–287.
11. Adrian J. Jervis, Jonathan A. Butler, Andrew J. Lawson, Rebecca Langdon, Brendan W. Wren, and Dennis Linton, "Characterization of the structurally diverse N-linked glycans of Campylobacter species," *Journal of Bacteriology*, vol. 194, no. 9, pp. 2355–2362, 2012.

Chapter 12

Artificial Intelligence and Machine Learning Techniques in Power Systems Automation

Amar Nayak and Rachana Kamble

Contents

DOI: 10.1201/9781003301820-12

12.1 Introduction

Artificial intelligence (AI), machine learning (ML), and deep learning (DL) are being applied in a variety of ways in the power systems industry to improve efficiency, reliability, and overall performance [1]. Some examples include demand forecasting, distribution system management, fault detection and diagnosis, asset management, electricity market analysis and trading, power generation, and renewable integration. AI, ML, and DL can be used to enable self-healing grid, enhance efficiency, and reduce operational costs in various power system applications such as distribution grid, transmission grid, renewable energy integration, and power plants. There are many more possibilities and advancements happening in power systems in the research field continuously due to AI, ML, and DL.

With the advent of IoT and cheap data storage, large data sets will be available and AI/ML will play a crucial role in turning that data into valuable insights for power system engineers and operators [2]. ML methods have better robustness against different systems, better adaption to system uncertainties, and less dependent on the modeling accuracy and validity of assumptions.

The chapter is organized as follows. Section 2 provides a background of AI, ML, and DL. Section 3 presents a literature review related to AI, ML, and DL applications in power systems. Section 4 discusses the applications of artificial intelligence, machine learning, and deep learning in power systems. Section 5 provides the parameters used by AI, ML, and DL algorithms for analysis of power systems generation, translation, and distribution process. Section 6 presents the conclusions.

12.2 Background

12.2.1 Electric Power Systems

Electric power systems refer to the infrastructure and equipment that is used to generate, transmit, distribute, and control the flow of electrical energy. This includes the power plants where electricity is generated, the transmission and distribution lines that transport the electricity over long distances, and the substations and transformers that help to regulate and control the flow of electricity.

12.2.2 Electric Power Systems Have Several Main Components

Generation: This includes all the power plants, such as coal, gas, nuclear, and renewable energy sources, where electricity is generated.

Transmission: The transmission system is the backbone of the electric power system, comprising high-voltage power lines and substations that transport electricity over long distances.

Distribution: The distribution system is the last step in the delivery of electricity to customers. It includes the lower-voltage power lines, transformers, and substations that bring the electricity from the transmission system to the end-users.

Control and Monitoring: This includes the systems and technologies that are used to monitor and control the operation of the power system, such as supervisory control and data acquisition (SCADA) systems, energy management systems (EMSs), and advanced metering infrastructure (AMI) systems [3].

The electric power system is a complex network of generators, transmission and distribution lines, transformers and substations, and various control and monitoring systems. The main goal of electric power systems is to produce, transmit, distribute, and manage the electric energy in a secure, reliable, efficient, and economical way to meet the needs of customers.

Electric power systems are constantly evolving as new technologies, such as renewable energy sources, energy storage systems, and advanced grid control systems, are developed and integrated into the existing infrastructure.

12.2.3 Electric Power Systems Structure

The structure of a power system typically includes three main parts: the generation sector, the transmission sector, and the distribution sector. See Figure 12.1.

Generation sector: This sector includes power plants that generate electricity using various sources of energy such as coal, natural gas, nuclear, hydro, wind, and solar. The generated electricity is then fed into the transmission grid.

Transmission sector: The transmission sector is responsible for moving electricity from the generation sites to the areas where it is needed. This sector includes high-voltage power lines, substations, and transformers [4]. The electricity is transported over long distances through these high-voltage transmission lines and is then stepped down to lower-voltage levels at substations, to be ready for distribution.

Distribution sector: The distribution sector is responsible for delivering electricity to the end-users, such as homes, businesses, and industrial customers. This sector includes a network of lower-voltage power lines, substations, and transformers.

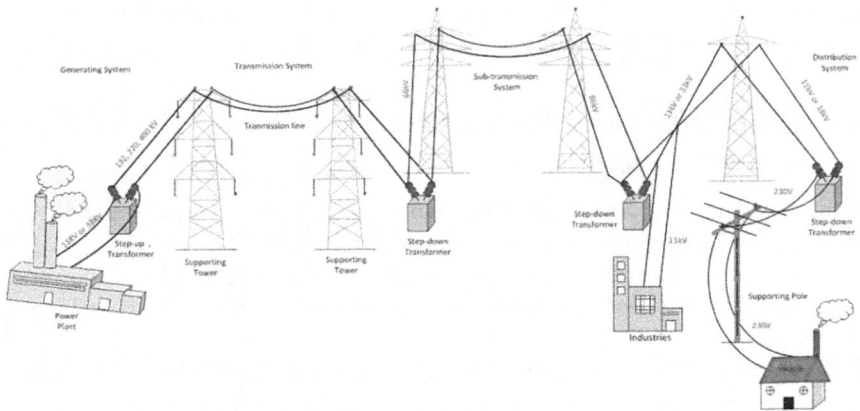

Figure 12.1   Structure of Power System.

The electricity is distributed to customers through these lower-voltage lines and is regulated to the appropriate voltage level at substations.

The power system is connected by a combination of overhead power lines, underground power cables, and substations. The power system is constantly monitored and controlled by various control systems, such as SCADA systems, EMSs, and AMI systems.

Additionally, the power system is equipped with protection and control systems to ensure safe and reliable operation of the system and protection of the equipment. These include protection relays, fault current limiters, and power electronics devices like static var compensator, static synchronous compensator, etc.

Overall, the structure of a power system is designed to ensure that electricity is produced, transmitted, distributed, and used efficiently and reliably to meet the needs of customers.

12.2.4 *Introduction to Artificial Intelligence (AI)*

Artificial intelligence (AI) refers to a field with the combination of computer science and robust data sets that is capable of solving problems. It is also enhancing the simulation of human intelligence in machines that are programmed to think and learn like individuals. AI systems can be designed to perform a wide variety of tasks, including perception, reasoning, decision-making, natural language understanding, and learning [5].

There are different types of AI, each with its own set of capabilities and limitations. Some examples include the following:

Rule-based systems: These systems are based on a set of predefined rules and can perform tasks such as simple decision-making and data validation.

Expert systems: These systems are designed to mimic the decision-making abilities of a human expert in a specific field and can be used for tasks such as medical diagnosis or financial forecasting.

Neural networks: These systems are inspired by the structure of the human brain and are composed of interconnected nodes, or "neurons," that can learn from data. Neural networks can be used for tasks such as image recognition, natural language processing, and speech recognition.

Deep learning: A subset of machine learning, which is based on neural networks with multiple layers, it can learn from raw and unstructured data and can be used for performing complex tasks, such as speech and image recognition.

Reinforcement learning: This type of AI involves training an agent to make decisions in an environment by rewarding it when it makes good decisions and punishing it when it makes bad decisions. This approach is used in complex decision-making tasks, such as robotics, video games, and finance.

Overall, the field of AI is constantly evolving, and new techniques and approaches are being developed all the time. Some experts believe that AI has the potential to revolutionize a wide range of industries, from healthcare to transportation, and will play an increasingly important role in our daily lives.

12.2.5 Introduction to Machine Learning

Machine learning is a subset of artificial intelligence (AI) that involves the development of algorithms and statistical models that enable computers to learn from and make decisions or predictions based on collected data; there is no need to explicitly program to do so. The key idea behind machine learning is that these algorithms and models can improve their performance through experience by learning from historical data [6].

There are several types of machine learning, each with its own set of techniques and approaches. Some examples are as follows:

Supervised learning: This type of machine learning involves training a model on a labeled data set, where the desired output is already known. The model can then be used to make predictions on new, unseen data. Examples of supervised learning tasks include classification and regression problems.

Unsupervised learning: This type of machine learning involves training a model on an unlabeled data set, where the desired output is not known. The goal of unsupervised learning is to identify patterns or structures in the data. Dimensionality reduction and clustering are the examples of unsupervised learning tasks.

Semi-supervised learning: This type of machine learning is a combination of supervised and unsupervised learning. It's used when there is a limited amount of labeled data available. The model can use the labeled data to make predictions and the unlabeled data to find patterns and features in the data.

Reinforcement learning: This type of machine learning involves training an agent to make decisions in an environment by rewarding it when it makes good decisions

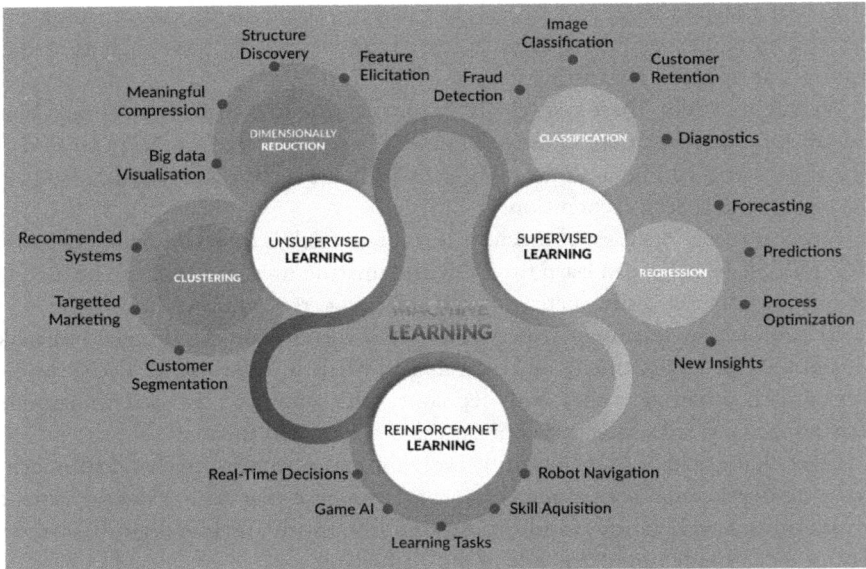

Figure 12.2   Machine Learning Applications.

and punishing it when it makes bad decisions. This approach is used in complex decision-making tasks, such as robotics, video games, and finance.

Machine learning algorithms have a wide variety of applications, including image recognition, natural language processing, speech recognition, finance, healthcare, marketing, transportation, and many more (see Figure 12.2). The algorithms can be run on large data sets and improve their decision-making capability as more data is fed in, allowing for continuous learning and adaptation.

12.2.5.1 *Machine Learning Applications*

- Self-driving car (Google)
- Virtual and augmented reality systems (Microsoft HoloLens, HTC Vive)
- Personal assistants (Google Now, Siri, Cortana)
- Social media services (Facebook, Instagram)
- Product recommendation (Amazon, Flipkart, Netflix, YouTube)
- Power systems generations, transmission, and distributions
- Chatbots (Mitsuku, Replica)
- Spam filtering (Gmail)
- Search engines (Google, Bing)
- Predictive analysis (order return prediction, sales prediction)
- Anomaly detection (fraudulent transaction, behavior analysis)

- Various learning paradigms
- Education
- Transport
- Service and home robots
- Public safety and security
- Healthcare
- Entertainment
- Low resource communication
- Employment and workspace

12.2.6 Introduction to Deep Learning

Deep learning is a subset of machine learning that is based on artificial neural networks with many layers (hence "deep") [7]. These networks are designed to learn from large and complex data sets and are particularly well-suited for tasks such as image recognition, natural language processing, and speech recognition.

A neural network is a mathematical model inspired by the structure of the human brain that is composed of interconnected nodes, or "neurons." These neurons are organized into layers, where the input layer receives the data, and the output layer produces the prediction or decision. Between these layers, there are one or more hidden layers, which process and analyze the data. The more hidden layers a network has, the "deeper" the network is, thus the name deep learning.

Deep learning algorithms are trained using a process called backpropagation, which involves adjusting the weights of the neurons in the network so that the output of the network is as close as possible to the desired output. The process is repeated many times with different training data, allowing the network to learn from the data and improve its performance.

Deep learning algorithms can be used for many tasks including image recognition, natural language processing, speech recognition, machine translation, drug discovery, and many more. As the technology of deep learning has matured and become more accessible, it is increasingly being applied to solve problems across different domains. It has become particularly useful for tasks that involve large and complex data sets such as image and speech recognition, natural language processing, and even anomaly detection, in which traditional machine learning models have struggled.

12.3 Literature Review

There has been a significant amount of research in the areas of AI, ML, and DL applied to power systems in recent years. Here are a few key areas where these techniques have been applied and some examples of relevant studies:

Demand forecasting: Many studies have applied ML and DL techniques to predict future electricity demand. An example: Liu et al. [8] used a deep neural network to predict electricity demand in China and found that it outperformed traditional statistical methods. Hoa et al. [9] applied a hybrid ML approach to predict electricity demand in Ontario, Canada, and found that it provided more accurate predictions than traditional time-series models.

Distribution system management: AI and ML have been applied in various ways to optimize the operation of distribution systems. Stock et al. [10] used a combination of swarm intelligence and ML to optimize the operation of distribution systems and found that it improved the overall performance of the system. Another study by Alabdullaha et al. [11] used a deep Q-network reinforcement learning to improve the control of distributed energy resources connected to distribution systems and found that it effectively reduced the amount of power lost in the system.

Fault detection and diagnosis: AI and ML have been widely used in the field of fault detection and diagnosis in power systems. An example: Wang et al. [12] proposed a deep convolutional neural network-based method for the detection and classification of power system transient faults, which was tested on a large data set and found to have high accuracy rate. Another study by Zhang et al. [13] used an improved artificial bee colony algorithm to diagnose power transformer faults and found that it improved the accuracy of fault diagnosis compared to traditional methods.

Asset management: AI and ML have also been applied to improve the maintenance and management of power system assets. A study by Nguyen et al. [14] used an ML-based method to predict the remaining useful life of power transformers and found that it could accurately predict when equipment is likely to fail. Another study by Li et al. [15] used an ML-based method for the condition monitoring of power system equipment and found that it improved the accuracy of equipment fault detection.

Electricity market analysis and trading: AI and ML have been used to analyze and predict electricity prices in various electricity markets. A study by Castelli et al. [16] proposed an ML-based method for predicting electricity prices in the day-ahead electricity market and found that it improved the accuracy of price predictions compared to traditional methods. Another study by Li et al. [17] used a LTRCNNM DL-based method for predicting prices in the real-time electricity market and found that it provided more accurate predictions than traditional time-series models.

Power generation and renewable integration: AI and ML have been applied in various ways to optimize the operation of power plants and to enable better integration of renewable energy sources into the grid. A study by Wong et al. [18] proposed an ML-based method for optimizing the operation of a microgrid, which was tested on a real-world system and found to improve the overall performance of the system. Another study by Yang et al. [19] used an ML-based method for real-time control of a wind farm and found that it effectively reduced the amount of power lost in the system.

These are some of the examples of the ongoing research activities in the field of AI, ML and DL applied to power systems. There are many more possibilities and advancements happening in the research field continuously.

12.4 Applications of Artificial Intelligence, Machine Learning, and Deep Learning in Power Systems

12.4.1 Applications of AI and ML in Power Systems

There are several areas in which artificial intelligence (AI), machine learning (ML), and deep learning can be used in electric power systems to improve their performance and reliability. Some examples include:

Predictive maintenance: AI and ML can be used to analyze data from power system equipment, such as transformers and transmission lines, to identify patterns and anomalies that indicate the equipment is approaching failure. This can enable utilities to schedule maintenance before equipment failure occurs, reducing the likelihood of power outages.

Smart grid management: AI and ML can be used to optimize the operation of the electric grid, such as by balancing supply and demand in real-time, to improve the efficiency and reliability of the grid. This can help utilities to reduce costs and improve the integration of renewable energy sources into the grid.

Power demand forecasting: Machine learning models can be trained on historical power demand data to make predictions about future power demand. These predictions can then be used to improve the operation of the power system, for example, by scheduling the operation of power plants and power transmission lines to meet anticipated demand.

Power quality monitoring: AI/ML-based algorithms can be applied for power quality monitoring; they can detect and classify power quality events in real-time, such as voltage sags, swells, and harmonic distortions, which can help to improve the reliability and efficiency of the power system.

Fault diagnosis and protection: AI and ML algorithms can be used to analyze data from power system sensors to detect and diagnose faults, such as short circuits, in real-time. This can improve the speed and accuracy of fault detection and isolation, which can help to reduce the likelihood of power outages and damage to equipment.

Renewable energy integration: Machine learning algorithm can be used to predict the output of solar and wind energy generation so that energy generation can be balanced with the energy demands on real-time basis.

These are just a few examples of how AI and ML can be applied to electric power systems. With the growth of IoT and data analytics, the opportunities for AI and ML in power systems are expected to expand and we'll see more advanced and dynamic applications in near future.

We have seen tremendous applications of AI and ML in power systems like fault detection, voltage control, planning, distribution analysis, scheduling, forecast, etc.

AI and ML can be used in power systems data analytics to extract valuable insights from large and complex data sets. Some examples of how AI and ML can be used in this context include the following:

Predictive modeling: AI and ML algorithms, such as neural networks and decision trees, can be used to create predictive models that can forecast future events, such as power demand, equipment failures, or grid disturbances. These models can be trained on historical data and can be used to make decisions that can improve the performance and reliability of the power system.

Anomaly detection: AI and ML algorithms, such as clustering and outlier detection, can be used to identify unusual or abnormal patterns in power system data. This can help utilities to detect and diagnose problems early, such as equipment failures or power theft, and can enable them to take corrective action before the problem becomes serious.

12.4.2 Applications of Machine Learning in Power System Analysis and Data Visualization

AI and ML algorithms can be used to visualize large and complex power system data sets, making it easier for operators and engineers to understand and interpret the data. This can help to identify trends and patterns in the data that might not be immediately apparent and can enable utilities to make data-driven decisions.

Automated reporting: AI and ML algorithms can be used to automatically generate reports from power system data. These reports can be used to provide real-time visibility into the performance of the power system and can be used to identify patterns and trends that can help utilities to improve their performance.

Process optimization: Machine learning models can be used to optimize the power system processes, such as generation, transmission, and distribution, by predicting the load on different areas of grid and adjusting the available power accordingly.

Root cause analysis: Machine learning models can be used to identify the root cause of power system incidents or equipment failure by analyzing the historical data from various sensors, including weather data, voltage, current, and temperature.

In summary, AI and ML can be used in power systems data analytics to analyze and interpret large and complex data sets in order to improve the performance and reliability of the power system. With the advent of IoT and cheap data storage, large data sets will be available, and AI/ML will play a crucial role in turning that data into valuable insights for power system engineers and operators.

12.4.3 Applications of Deep Learning in Power Systems

Deep learning, a subset of machine learning, can be used in a variety of ways in power systems to improve performance and reliability. Some examples of how deep learning can be used include:

Predictive maintenance: Deep learning algorithms, such as convolutional neural networks (CNNs) and recurrent neural networks (RNNs), can be used to analyze sensor data from power system equipment to predict when equipment is likely to fail. This can enable utilities to schedule maintenance before failure occurs, reducing the likelihood of power outages.

Grid control and optimization: Deep learning algorithms can be used to optimize the operation of the electric grid, such as by balancing supply and demand in real-time, to improve the efficiency and reliability of the grid. This can help utilities to reduce costs and improve the integration of renewable energy sources into the grid.

Fault detection and diagnosis: Deep learning algorithms, such as auto encoders and deep belief networks, can be used to analyze sensor data from power system equipment to detect and diagnose faults, such as short circuits, in real-time. This can improve the speed and accuracy of fault detection and isolation, which can help to reduce the likelihood of power outages and damage to equipment.

Renewable energy integration: Deep learning algorithms, such as deep neural networks, can be used to predict the output of renewable energy sources, such as solar and wind power, in order to optimize the integration of these sources into the power system.

Power demand forecasting: Deep learning models, such as long short-term memory (LSTM) networks, can be trained on historical power demand data to make predictions about future power demand. These predictions can then be used to improve the operation of the power system, for example, by scheduling the operation of power plants and power transmission lines to meet anticipated demand.

Power quality monitoring: Deep learning algorithms, such as CNNs and RNNs, can be used for power quality monitoring; they can detect and classify power quality events in real-time, such as voltage sags, swells, and harmonic distortions, which can help to improve the reliability and efficiency of the power system.

Deep learning models are particularly well suited for power systems applications, because they can learn to identify patterns and make predictions from large, complex, and non-linear data sets. With the vast amount of data available for power systems, deep learning models can be a very powerful tool.

12.5 Parameters Used by AI, ML, and DL Algorithms

12.5.1 Parameters Used by ML Algorithms in Power Systems Generation and Transmission

There are a variety of parameters that can be used as inputs for machine learning algorithms in power systems. Some examples include the following:

Power demand: Power demand is one of the most important parameters in power systems, as it is used to determine the amount of power that needs to be generated, transmitted, and distributed to meet the needs of customers. Power demand can be measured in terms of energy (e.g., kilowatt-hours) or power (e.g., megawatts).

Power generation: Power generation data can be used to train machine learning algorithms to predict the output of power plants, including renewable energy sources such as solar and wind power, to optimize the power system operation.

Power transmission and distribution: Machine learning algorithms can use data on the operation of power transmission and distribution systems, including voltage and current measurements, to improve the efficiency and reliability of the grid.

Weather data: Weather data, such as temperature, humidity, and precipitation, can be used as inputs for machine learning algorithms to predict the output of renewable energy sources and to optimize the operation of the power system.

Power quality: Machine learning algorithms can use data on power quality parameters, such as voltage and frequency, to detect and diagnose power quality events, such as voltage sags and swells, which can help to improve the reliability and efficiency of the power system.

Equipment data: Data from sensors on power system equipment, such as transformers, generators, and transmission lines, can be used to train machine learning algorithms to predict when equipment is likely to fail, which can help to improve the reliability of the power system.

Grid status and events: Machine learning algorithms can be trained on historical data of grid status and events such as grid congestion, branch outages, and load shedding, to identify patterns and potential vulnerabilities in the power grid, which can be used to improve the grid's stability and reliability.

These are just a few examples of the parameters that can be used as inputs for machine learning algorithms in power systems. The specific parameters used will depend on the application and the goals of the machine learning model.

12.5.2 Parameters Used by AI, ML, and DL Algorithms in Smart Grids

Several parameters can be used as inputs for machine learning algorithms in smart grid applications in power systems. Some examples include the following:

Power demand: Power demand data can be used by machine learning algorithms to predict future power demand, which can help utilities to optimize the operation of power generation, transmission, and distribution systems to meet anticipated demand.

Power generation: Power generation data, including data on renewable energy sources such as solar and wind power, can be used by machine learning algorithms to optimize the integration of these sources into the grid.

Power transmission and distribution: Machine learning algorithms can use data on the operation of power transmission and distribution systems, such as voltage and current measurements, to optimize the operation of the grid and improve its efficiency and reliability.

Weather data: Weather data, such as temperature, humidity, and precipitation, can be used to predict the output of renewable energy sources and to optimize the operation of the grid in light of changing weather conditions.

Power quality: Machine learning algorithms can use data on power quality parameters, such as voltage and frequency, to detect and diagnose power quality events, such as voltage sags and swells, which can help to improve the reliability and efficiency of the grid.

Battery storage: Machine learning algorithms can use data from battery storage system such as charge/discharge, capacity, and state of charge to optimize the operation of these systems and enhance grid stability and reliability.

Smart meter data: Machine learning algorithms can use data from smart meters, such as energy consumption and production data, to optimize the operation of the grid and to identify patterns of energy use that can be used to improve energy efficiency.

Grid events data: Machine learning algorithms can use data on historical grid events, such as outages and congestion, to identify patterns and potential vulnerabilities in the grid and to improve its stability and reliability.

These are just a few examples of the parameters that can be used as inputs for machine learning algorithms in smart grid applications in power systems. The specific parameters used will depend on the application and the goals of the machine learning model.

12.5.3 Parameters Used by AI, ML, and DL Algorithms in Cyber Security

There are several parameters that can be used as inputs for machine learning algorithms in the context of cyber security for power systems. Some examples include the following:

Network traffic data: Machine learning algorithms can analyze network traffic data to identify patterns and anomalies that may indicate a cyber attack. This can include data on network connections, packet headers, and payloads.

System logs: Machine learning algorithms can analyze system logs, such as those generated by firewalls and intrusion detection systems, to detect signs of a cyber attack. This can include data on failed login attempts, system crashes, and unusual process behavior.

User behavior data: Machine learning algorithms can analyze data on user behavior, such as keystroke dynamics, mouse movements, and login patterns, to detect signs of a cyber attack.

SCADA and ICS data: Machine learning algorithms can analyze data from supervisory control and data acquisition (SCADA) and industrial control systems (ICSs) to detect abnormal patterns that may indicate a cyber attack.

Physical sensor data: Machine learning algorithms can analyze data from physical sensors, such as temperature and vibration sensors, to detect signs of a cyber attack. For example, a sudden increase in temperature in certain equipment might be a sign of a malware attack on that device, turning it into a malware distribution point.

Historical event data: Machine learning algorithms can be trained on historical data of events, such as successful and attempted cyber attacks, to identify patterns and potential vulnerabilities in the power system.

By analyzing these parameters, machine learning algorithms can identify patterns and anomalies that may indicate a cyber attack and trigger an alert for system operators to take action. Additionally, by leveraging historical data, machine learning algorithms can continuously improve the detection of potential security threats.

12.6 Conclusions

There are several areas in which artificial intelligence (AI) and machine learning (ML) can be applied to electric power systems to improve their performance and reliability. ML applications have seen tremendous adoption in power system research and applications. This chapter discussed how AI and ML can be used in power systems data analytics to analyze and interpret large and complex data sets in order to improve the performance and reliability of the power system. By analyzing power generation, weather, power quality, power transmission and distribution, smart meter, grid events, and battery storage data, machine learning algorithms can improve the performance and reliability of the power system. By analyzing physical sensor, SCADA and ICS, historical event, user behavior, system logs, and network traffic data parameters, deep learning algorithms can identify patterns and anomalies that may indicate a cyber attack and trigger an alert for system operators to take action. Additionally, by leveraging historical data, machine learning algorithms can continuously improve the detection of potential security threats.

References

1. M.S. Ibrahim, W. Dong, and Q. Yang, "Machine learning driven smart electric power systems: Current trends and new perspectives", *Appl. Energy*. 2020, 272 (February), 115237. http://dx.doi.org/10.1016/j.apenergy.2020.115237.
2. M. Farhoumandi, Q. Zhou, and M. Shahidehpour, "A review of machine learning applications in IoT-integrated modern power systems", *Electr. J.* 2021, 34 (1), 106879. http://dx.doi.org/10.1016/j.tej.2020.106879.
3. Rajeev Kumar, M.L. Dewal, and Kalpana Saini, "Utility of SCADA in power generation and distribution system", International Conference on Computer Science and Information Technology, IEEE. 2010, 297–306.
4. B. Jimada-Ojuolape and J. Teh, "Impact of the integration of information and communication technology on power system reliability: A review", *IEEE Access*. 2020a, 8, 24600–24615. http://dx.doi.org/10.1109/ACCESS.2020.2970598.

5. Z. Shi et al., "Artificial intelligence techniques for stability analysis and control in smart grids: Methodologies, applications, challenges and future directions", *Appl. Energy.* 2020, 278 (July), 115733. http://dx.doi.org/10.1016/j.apenergy.2020.115733.
6. E.U. Haq, X. Lyu, Y. Jia, M. Hua, and F. Ahmad, "Forecasting household electric appliances consumption and peak demand based on hybrid machine learning approach", *Energy Rep.* 2020, 6, 1099–1105. http://dx.doi.org/10.1016/j.egyr.2020.11.071.
7. Weibo Liu, Zidong Wang, Xiaohui Liu, Nianyin Zeng, Yurong Liu, and Fuad E. Alsaadi, "A survey of deep neural network architectures and their applications", *IEEE.* 2019, 23–56.
8. Ningning Liu, Ping Jiang, Ranran Li, and Yuyang Gao, "A novel composite electricity demand forecasting framework by data processing and optimized support vector machine", *Elsevier.* 2020, 260 (15) (February), 223–234.
9. Dinh Hoa Nguyen and Anh Tung Nguyen, "A machine learning-based approach for the prediction of electricity consumption", *IEEE.* 2019, 380–387.
10. Simon Stock, Davood Babazadeh, and Christian Becker, "Applications of artificial intelligence in distribution power system operation", *IEEE.* 2021, 9, 150098–150121.
11. Mohammed H. Alabdullaha and Mohammad A. Abido, "Microgrid energy management using deep Q-network reinforcement learning", *Elsevier.* 2022, 61 (11) (November), 9069–9078.
12. Yixing Wang, Meiqin Liu, and Zhejing Bao, "Deep learning convolutional neural network for power system fault diagnosis", *Chinese Control Conference (CCC).* 2016. DOI:10.1109/ChiCC.2016.7554408.
13. Yiyi Zhang, Hongbo Peng, Jiake Fang, Liuliang Zhao, Xin Li, and Changyi Liao, "Transformer fault diagnosis based on new features selection and artificial bee colony optimization SVM", International Conference on Power System Technology (POWERCON), IEEE. 2018.
14. K. Nguyen, Sungin Cho, and J. Wetzer, "Conceptual design for asset management system under the framework of ISO 55000", *IEEE Int Conf. Electr Distrib* 2015, 1, 17–21.
15. Shengtao Li, and Jianying Li, "Condition monitoring and diagnosis of power equipment: Review and prospective", *IET.* 2017, 2 (2), 82–89.
16. Mauro Castelli, Aleš Groznik, and Aleš Popovič, "Forecasting electricity prices: A machine learning approach", *MDPI, Algorithms.* 2020, 13 (119), 1–16. DOI:10.3390/a13050119.
17. Xingpeng Li, Vasudharini Sridharan, and Mingjian Tuo, "Wholesale electricity price forecasting using integrated long-term recurrent convolutional network model", *MDPI, Energies.* 2022, 15, 7606.
18. Rohit Trivedi and Shafi Khadem, "Implementation of artificial intelligence techniques in microgrid control environment: Current progress and future scopes", *Elsevier, Energy and AI.* 2022, 8 (May), 100147.
19. Ru Li, Jincheng Zhang, and Xiaowei Zhao, "Dynamic wind farm wake modeling based on a bilateral convolutional neural network and high-fidelity LES data", *Elsevier, Energy.* 2022, 258 (1) (November), 124845.

Index

Note: Page numbers in *italics* indicate a figure on the corresponding page.

For Product Safety Concerns and Information please contact our EU
representative GPSR@taylorandfrancis.com
Taylor & Francis Verlag GmbH, Kaufingerstraße 24, 80331 München, Germany

www.ingramcontent.com/pod-product-compliance
Lightning Source LLC
Chambersburg PA
CBHW060407220326
41598CB00023B/3050